STORM WARNING

STORM WARNING

THE ORIGINS OF THE WEATHER FORECAST

PAULINE HALFORD

FOREWORD BY
MICHAEL FISH MBE

SUTTON PUBLISHING

First published in the United Kingdom in 2004 by
Sutton Publishing Limited · Phoenix Mill
Thrupp · Stroud · Gloucestershire · GL5 2BU

Reprinted 2005

British Library Cataloguing in Publication Data
A catalogue record for this book is available from the British Library.

ISBN 0-7509-3215-5

Typeset in 10.5/12.5pt Melior.
Typesetting and origination by
Sutton Publishing Limited.
Printed and bound in England by
J.H. Haynes & Co. Ltd, Sparkford.

Contents

List of Illustrations

(Between pages 126 and 127)

Foreword

What better book to commemorate our 'Year of Weather'?

2004 has seen the BBC celebrate fifty years of live TV weather forecasts and announce plans for a brand new style of presentation. The Met Office reached its 150th anniversary – and, subsequently, the world's first Meteorological Service opened its new headquarters in Exeter. And, on a personal note, 2004 marked my thirtieth anniversary of TV broadcasting (there can't be many broadcast meteorologists in the world to have lasted that long) and my retirement after forty-two years at the Meteorological Office.

This book takes us on a journey from the ancient Greeks and Romans, when the climate was controlled by the Gods, through to the modern day. And, to be honest, there was little advancement in the science of meteorology for the two millennia following those ancient societies until the invention of the barometer and thermometer. The journey takes us from those humble beginnings during the time of the Crimean War, through the late 1950s when computers and satellites came into play, to the modern day with the Met Office that is home to one of the most powerful computers in the world. It is fed by data from satellites, ships, planes, buoys, balloons, radar and even the occasional human being.

We are also offered a wonderful insight into the life and work of the founding father of the Met Office, Admiral FitzRoy. He suffered from the frustrations of Civil Service bureaucracy, rivalry and workplace jealousy, which eventually led to his suicide. At least the modern-day Met Office employees are issued with a thick skin suit on their arrival!

FitzRoy never had the slightest inkling of global warming – now one of the greatest threats to mankind. And he would be amazed at the meteorological advancements of the last few years. A wrong forecast is a rarity, whereas in his day a correct one was virtually unheard of. One hardly dares to predict the progressions we shall see in the next 150 years.

I have a lot to thank him and the conference in Brussels of 1853 for. Between them they committed the British government to form the first Meteorological Office – without them, I would not have had a wonderful and fulfilling forty-two-year career utilising the other great legacies from that period: storm warnings, synoptic charts, worldwide observations and press forecasts.

Fittingly, I end with the words of Admiral FitzRoy himself: 'the problem with the weather is that the only certain thing about it is its unpredictability.'

Michael Fish, MBE

Acknowledgements

Had I known what a vast and ambitious venture I had set out upon with this book, perhaps I would never have tried. However, I was naive enough to think that my task was achievable, so, like my heroes Robert FitzRoy, Matthew Maury, Christoph Buys Ballot, the Shepherd of Banbury and all who strove with the even more daunting enterprise of attempting to foretell the weather, I battled on in the belief that the result would be worthy of the effort. Whether I have succeeded anywhere near as well as they did is not for me to say. I can say, though, that unlike them, I have not been subjected to ridicule, disbelief and antagonism while I laboured, only help and encouragement from many people, all of whom merit my sincere thanks.

To list some among the many, they are: Bernard Ashley; the British Association (Ben Savage and Lucy Johnson); the British Library; Buckinghamshire County Reference Library; Illustrated London News Picture Library (Katie Simpson); Meteorological Office; National Archives, Kew; National Maritime Museum, Greenwich (Lucy Waite); Royal Geographical Society; Royal Meteorological Society History Group; Royal Society Library; Science Museum; Science and Society Picture Library (Susie Kocher); Smithsonian Institution, Washington; Westminster Archives; Ken Woodley.

I would like to give a special mention to Steve Jebson, Graham Bartlett and all the staff of the Met Office's National Meteorological Library and Archives, who could not have been more welcoming and helpful in the many weeks I spent with them in Bracknell – and I am especially grateful to Graham, who cheerfully extracted old and fragile tomes from boxes packed ready for the move of the Library and Archives from Bracknell to Exeter; to Derek Barlow, whose magnifice catalogue of the correspondence and papers relating to the fi*t Meteorological Office guided me unerringly through *ne mountain of documents held in the National Archives; ar* to meteorologist John Hunter, who lent his scientific knowledge and skills to this project – though I stress that any errors and omissions are entirely my own. I must not forget, either, Mrs Richmond-Watson, who conducted me on an extensive and thoroughly absorbing tour of Wakefield Lodge, Potterspury, the former residence of the FitzRoy family and now her own home; and Dr Howard Oliver, whose lecture to the Royal Meteorological Society History Group added much to my knowledge of John Dalton.

To this list I would like to add the many friends who have given of their time and advice throughout months of research and writing, Jaqueline Mitchell and everyone at Sutton Publishing, and above all my long-suffering husband, Mike, without whose support and understanding this book simply would not have happened.

Pauline Halford

Principal Characters

George Biddell Airy Astronomer Royal, 1835–81. President of the Royal Society, 1871–3.

Thomas Henry Babington Junior clerk at the Meteorological Office from its foundation in 1854. Became acting head of the Office from FitzRoy's death in 1865 until storm warnings ceased in 1866.

Alexander Dallas Bache Superintendent of the US Coastal Survey from 1843. Close friend and supporter of Joseph Henry.

Sir Francis Beaufort Hydrographer of the Royal Navy until 1857 and inventor of the Beaufort scale of wind speeds in 1805.

Captain Frederick William Beechey Arctic explorer and naval surveyor who became Superintendent of the Marine Department of the Board of Trade from its founding in 1850. Represented Britain at the Brussels Conference of 1853.

Major-General Sir John Burgoyne Inspector of Fortifications, Royal Engineers. Authorised a worldwide chain of meteorological observation stations in 1849.

Christoph Buys Ballot Dutch meteorologist. Founded the Dutch Royal Meteorological Institute, 1854. Introduced a system of storm warnings in 1860 and daily weather forecasting in the Netherlands.

Henrich Dové Director of the Royal Prussian Meteorological Institute and writer of the definitive work *The Law of Storms*, published in 1857.

James Pollard Espy American former teacher. In the 1830s developed a theory of storms based on the upward motion of air at the centre.

Thomas Farrer Assistant Secretary to the Marine Department of the Board of Trade, 1850. Promoted to Assistant Secretary to the Board of Trade, 1854, then Permanent Secretary, 1865–86.

Vice-Admiral Robert FitzRoy Former captain of the *Beagle*. Founder of the Meteorological Office in 1854 and inventor of the weather forecast in 1861.

Francis Galton Cousin of Charles Darwin. Leading member of the Meteorological Committee of the Royal Society, 1867–77, and of the Meteorological Council, 1877–1901.

James Glaisher Meteorological Officer at the Greenwich Observatory under Airy and Secretary of the Meteorological Society from its founding in 1850.

Joseph Henry Head of the Smithsonian Institution in Washington, from its founding in 1846.

Sir John Herschel Son of astronomer Sir William Herschel. Made significant observations of Halley's comet in South Africa in 1835. A leading all-round scientist of his time.

Luke Howard Biologist and chemist, His *Essay on the Modifications of Clouds* (1804) first gave names to the different types of clouds.

Capt Henry James Superintendent of the Royal Engineers' meteorological observations. Represented Britain at the Brussels Conference of 1853.

Jean-Baptiste Antoine de Monnet, Chevalier de Lamarck French biologist and naturalist. Worked with Lavoisier in the setting up of a series of weather stations in France at the end of the eighteenth century.

Antoine Lavoisier French chemist. Worked with Lamarck in the setting-up of a series of weather stations in France at the end of the eighteenth century.

Urbain Le Verrier Director of the Imperial Observatory, Paris, 1853–69 and 1872–7. Joint discoverer (with John Adams) of the planet Neptune in 1846.

E.H. Marié-Davy Director of Magnetism and Meteorology at the Imperial Observatory, Paris, from 1862. Started the French storm warning service in 1863.

Capt Matthew Fontaine Maury Superintendent of the National Observatory in Washington. Convened the Brussels Conference of 1853.

Adolphe Quetelet Founded the Belgian Royal Observatory in 1833 and hosted the Brussels Conference of 1853.

William C. Redfield American steamship line owner. In the 1830s proposed a theory of storms based on a circular motion of wind around a low pressure centre.

Lt-Col William Reid Officer of the Royal Engineers. Contributed pioneering research into storms during the 1840s and 1850s.

Lt-Col Edward Sabine General Secretary of the British
  Association for the Advancement of Science, 1839–58.
  President of the Royal Society, 1861–71.
Robert H. Scott Head of the Meteorological Office, 1867–1900.
  Reintroduced the regular daily forecast in 1879.
Captain Bartholomew Sulivan Lieutenant with FitzRoy on the
  Beagle. Appointed superintendent of the Marine
  Department of the Board of Trade in 1857.
Lord John Wrottesley President of the Royal Society, 1854–8,
  and member of the Balloon Committee of the British
  Association.

SIGNS OF RAIN

by Dr Edward Jenner

The hollow winds begin to blow,
The clouds look black, the glass is low,
The soot falls down, the spaniels sleep
And spiders from their cobwebs peep.
Last night the sun went pale to bed,
The moon in halos hid her head;
The boding shepherd heaves a sigh,
For see! A rainbow spans the sky.
The walls are damp, the ditches smell,
Closed is the pink-eyed pimpernel.
Hark! How the chairs and tables crack.
Old Betty's joints are on the rack;
Her corns with shooting pains torment her,
And to her bed untimely send her.
Loud quack the ducks, the peacocks cry,
The distant hills are looking nigh.
How restless are the snorting swine!
The busy flies disturb the kine.
Low o'er the grass the swallow wings;
The cricket, too, how sharp he sings!
Puss on the hearth, with velvet paws,
Sits wiping o'er her whiskered jaws.
Through the clear stream the fishes rise,
And nimbly catch th' incautious flies.
The glow-worms, numerous and bright,
Illumed the woodland dell last night.
At dusk the squalid toad was seen
Hopping and crawling o'er the green.
The whirling dust the wind obeys,
And in the rapid eddy plays.
The frog has changed his yellow vest,
And in a russet coat is dressed.
Though June, the air is cold and still,
The mellow blackbird's note is shrill;
My dog, so altered in his taste,
Quits mutton bones on grass to feast.
And see, yon rooks, how odd their flight,
They imitate the gliding kite,
And seem precipitate to fall,
As if they felt the piercing ball –

'Twill surely rain – I see with sorrow
Our jaunt must be put off tomorrow.

Prologue

The waters of the Black Sea pounded against the cliffs at the mouth of Balaclava harbour. Cowering in his bunk aboard the *Retribution*, George, Duke of Cambridge, gibbered in fear of his life. One almighty crash had already had him screaming as a rudder was rent from the stern and now the second had gone. The ship, heavy with armament, listed violently and would surely capsize.

Out on deck, the captain snapped an order; the cannon were loosened and slid over the side into the foaming water. The ship, relieved of its top load, squared more happily into the gale. The Duke of Cambridge would survive to sail back to London to report his first-hand experiences of the Crimean campaign to his cousin, Her Majesty, Queen Victoria.

The storm had risen rapidly. 'It began at seven in the morning,' wrote Admiral Sir Leopold George Heath, KCB, 'and between eight and ten was at its height, the ships in harbour (I was on board *Sanspareil*) all drove on top of the other, but being moored head and stern they were all squeezed one on top of the other, side by side, and comparatively little damage was done to them'.

Outside the harbour, the rest of the fleet, which just two hours before had ridden proudly at anchor, was now scattered over a raging sea. Large transports, their hefty masts and spars snapped like so many matchsticks, smashed helplessly into smaller vessels and dashed themselves against the rocky shore. The Revd S. Kelson Stothert, in the Crimea as chaplain to the Naval Brigade, watched from on board the transport ship, *Queen*:

> During the squalls which every now and then blew through the rigging, some doomed ship snapped her cables and drifted helplessly on to the Cossack-lined beach . . . During a lull in the gale, Captain Mitchell obtained leave by signal to send boats to a wreck . . . A volunteer crew was quickly found for

three boats and, at great personal risk succeeded in bringing off a large portion of the frozen, half-starved crew, together with two women, wives of soldiers. The Cossacks fired at our boats, and a ball went through the bonnet of a woman and killed a Blue-jacket of ours, passing right through his head. We have saved 80 men, including officers, and the *Fury* 20. The crew of the *Lord Raglan* are all prisoners, with the lieutenant agent who was on board. The crew of a boat from the *Ville de Paris* was taken by the Cossacks.

Waves towered and plunged, studded with broken decking, tattered canvas, shredded ropes, bobbing barrels and drowning men. Clinging stubbornly to a raft of planks, semi-conscious and numb with cold, Midshipman Cosgrave was washed ashore near Balaclava. Just minutes earlier he had been standing on the deck of his ship helping cut away the mizzen mast at the order of his captain. The *Prince* was a huge, new, ironclad steam-screw transport and they thought it would withstand anything, which well it might have done, had not the loosened mizzen rigging fouled the screw. The ship careered out of control, was overtaken by the seas and sank. Of the crew of 150, only Cosgrave and a mere six men were saved.

Fortunately for the soldiers of the 46th (South Devonshire) Foot, they had disembarked from the *Prince* six days earlier at Balaclava: 'a most beautiful little harbour not more than 200 yards across, and cannot be seen from the sea,' reported Captain Nicholas Dunscombe, from the small steamer transferring him and his troops from ship to shore. However, his favourable impression did not last long. Through thick mud and driving rain, he marched his men the 8 miles from Balaclava to their encampment on the Heights of Sebastopol. At 4.30 p.m. on 13 November (the day before the storm) Captain Dunscombe and his men settled in the trenches for their first night patrol. Next day, in his diary he wrote:

It rained, hailed and snowed the whole day and night; besides all this it blew a perfect hurricane, the whole time I was there, I had nothing to eat or drink as on account of the severity of the weather our rations could not be sent to us; the men suffered so severely from cold, hunger, and wet and every privation that out of 160 that we took with us last night to the entrenchments, we only brought away 98 of them this evening; all the others

remained behind either dead, dying or so bad with the cramps
that they were not able to walk home . . .

'Home' was not a welcoming place to walk to, either. What
tents there were had been torn into tatters by the gale, tables
were broken, chairs scattered, barrels smashed and even the
sick and wounded lay open to the weather. Amidst the chaos,
Dunscombe was astounded to see a man he recognised come
staggering into the camp. He had last seen Midshipman
Cosgrave cheering and waving from the deck of the *Prince* as
he had disembarked. Cosgrave's sorry news, that most of the
crew had perished, was shocking enough, yet even more
crushing was the knowledge that with the *Prince* at the bottom
of the Black Sea lay the tents, provisions, medical supplies,
coats, gloves, socks and the 40,000 pairs of boots so badly
needed by the troops for the winter. 'If this weather continues
we certainly shall not be able to winter here, as the men even
at this early stage of the winter are dying in dozens,' remarked
Dunscombe, while the Revd Stothert, looking out from the
Queen over the remains of the French and British fleets, was
moved to declare: 'I could not help reflecting on the solemn
scene, and thinking how powerless we are with all our art and
and all our science.'

Back in England, many more were thinking the same. As
news of events in the Crimea began to appear in the columns
of the daily papers – the Revd Stothert himself reported to *The
Times* on the state of the ships and the sufferings of the troops
– people's patriotic pride turned to horror. But who, in 1854,
had sufficient climatic knowledge to realise that a Crimean
winter could be exceptionally wet and cold? And as for the
storm that devastated the naval fleets and sank the *Prince* with
its cargo of precious clothing and blankets, who could possibly
have foreseen that weather of such violence was about to strike
so suddenly?

There was a man who thought he could, and his name was
Captain Robert FitzRoy. FitzRoy was a naval captain of no
mean reputation – he had circumnavigated the globe in his
barque, the *Beagle*, and completed extensive surveys of the
notoriously stormy coasts around Cape Horn. His survival had
depended not only on his outstanding seamanship, but also on
his ability to read his weather glass and to interpret what it
was telling him about the weather to come. FitzRoy was not

stationed with the Fleet off Balaclava; his days in active naval service were long gone. He was, though, in a unique position to transform theory into action.

For 1854 not only saw the start of the Crimean War; it also saw the founding of Britain's first official Meteorological Office, and Captain (later Admiral) Robert FitzRoy, RN, was its very first chief. The Office's original role had nothing to do with weather prediction. In 1854 the concept was so outlandish as to be thought impossible. And as for a daily weather forecast? In November 1854 even Robert FitzRoy would have been incredulous of the possibility. Yet seven years later, in 1861, he would be issuing the world's first published daily forecast.

They were years fraught with conflict, to be followed by an even deeper and more desperate struggle against the ridicule of politicians, bureaucratic obstruction and the scorn of a sceptical scientific hierachy. It was a struggle that would eventually claim FitzRoy's life. He was, though, not the only pioneer treading the stormy path to weather prediction. In the teeth of gales of criticism, another naval officer, Lieutenant Matthew Maury of the US Navy, was striving to chart the winds, while, in Napoleonic France, Le Verrier and Marié-Davy of the Imperial Observatory fought each other in a battle to beat the British to a system of storm warnings.

From such turbulent beginnings were to emerge today's global forecasting services, with their panoply of space satellites, their networks of weather stations and an array of the world's most advanced computers. Despite all this technology, forecasts are still known to fail, and the public tends to regard weather prediction as much the same game of chance as it was for our ancestors. Yet forecasting is now an indispensible consitituent of modern life – aeroplanes will not take to the air without it, nor ships to sea, it is essential to commerce, and the military rely on it as a strategic weapon in the arena of war. General forecasts are published in every newspaper and broadcast twenty-four hours a day on dedicated television channels. We might scorn and deride them, but we still consult the weathermen – and women – rather than a piece of soggy seaweed before deciding whether or not to take a raincoat when we set out in the morning.

So how was it that an admiral of the British Navy became the very first weatherman? And why did it take a storm, and the destruction of a naval fleet, to trigger his transformation? The reason is simple. Forecasting was born of storm warnings, and it was because of the need of seamen to be forewarned of storms, because of naval exigency and all in the cause of safety at sea, that science came to be harnessed to foretell the weather. It was through the weather glass that science was first enlisted, but man's effort to predict the weather began way before the invention of the barometer. It has existed as long as man.

ONE

Myths to Meteorology

From the time Stone Age man first stuck his finger in the wind and declared 'it's going to rain today!', attempting to predict the weather has been a preoccupation of mankind. Like the animals, prehistoric man must have learnt to distinguish the seasons – winter was cold, summer hot, autumn a plenty of fruits and seeds, spring the time for fresh shoots, new plants and leaves. That is all very well on the grand scale of things, but it was the detail of the weather that controlled his existence on a daily basis. When he left his cave at daybreak, setting out to hunt for the day, where would be best to stalk his prey? Would they be basking by the water, or shading from the heat? Would gales and wind drive them to shelter or would they be grazing on the plains? Perhaps like the animals, prehistoric man reacted to an inner instinct that sensed the coming weather in the atmosphere, an instinct that civilisation has since stolen from us.

Even today, we attribute to animals, to birds and to nature an ability to predict the weather that we do not possess. How many of us see a herd of cows sitting in a field and take it as a sign of rain? Or take a good crop of red berries on the holly as a warning of a hard winter? And we all know, and like to believe, the old rhyme that says 'red sky at night, shepherd's delight'. Then there are the folk who predict the weather by remembering weather patterns from the past, or those who swear by the state of a piece of seaweed, or even those who see it foretold in their tea leaves. We tend to look on these predictions with indulgence, take note when they prove right and forgive and forget when they are proved wrong. When we do this, we are following a tradition handed down to us from prehistory, a tradition that was old when first documented by the ancient Greeks.

In Western culture it is the Greeks who provide the first
written evidence of a study of the weather. In the writings of
the ancient Egyptians weather features little, since, for them,
the weather held few mysteries. Their climate was largely
unchanging and their lives and harvests were dominated by
the Nile, whose yearly flood was almost as reliably predictable
as the rising of the sun itself. It was not until the peoples of the
Nile delta ventured out into the Mediterranean Sea that the
vagaries of the weather were of significance. For the Greeks,
though, as explorers of the seas, and with crops dependent on
the rains from the skies, the weather and its seasons held sway
over their lives.

Over the centuries, the Greeks identified a pattern in the
seasons that they codified in the 'Kalendar', an almanac in
which they divided the year into times suitable for agricultural
activities, such as ploughing, sowing and harvesting, and
times for building and for embarking on sea voyages. The
transition from one season to the next was marked by the
positions of the sun, the moon and the stars, and was the
occasion for a 'kalendar' festival, bequeathed to us today in the
timings of Christmas and Easter.

Although the eternal cycle of the heavens could mark the
seasons, it could not explain the nature of the weather. What
caused the winds and the storms? What brought drought and
flood? For this the Greeks looked to their gods – everyone
knew, for example, that thunderbolts were the work of Zeus,
the ruler of the heavens. Zeus was the god of the sky and he
resided in the ether, way above the earth. From there, this
omnipotent being controlled the atmosphere and everything in
it – the winds, the clouds, rain both wild and gentle, and all
the 'meteors', the extraordinary phenomena of the atmosphere.
Eos, or Aurora, was the shimmeringly beautiful goddess of
dawn who mated with Astraeus, the god of the starry sky, and
gave birth to the four winds – Boreas, the north wind,
Zephyrus the west wind, Eurus the east wind and Notus the
wind from the south. Boreas brewed storms and stirred the sea
into waves. Boreas' tempests were to be feared, especially by
mariners, yet even he could be relied upon to defend the ones
who worshipped him in their deepest hour of need. In 480 BC,
when all Athens trembled at the approach of Xerxes' vast
Persian fleet, the north wind rose and blew up a tremendous
storm that lasted three nights and two days. The Persians'

mighty triremes were sunk and scattered, and the small Athenian Navy triumphed in victory. But for Boreas, the tiny state of Athens would have been crushed by the might of the all-conquering Persians, and the grateful citizens built a splendid temple in his honour.

The Greek poet Aratus tells us, in his *Phaenomena*, that it is Zeus who formed the heavenly constellations. So the gods not only controlled the weather, they ruled the heavenly bodies as well, which in turn determined the seasons of the Kalendar. Thus the weather entered into the realm of religion – and woe betide any mortal committing the sacrilegious act of attempting to find any natural explanation for it, let alone trying to predict its behaviour.

This did not deter the mighty Greek philosophers, and it was they who made the first tentative scientific observations of weather phenomena and their relationships. At the end of the fifth century BC, the physician Hippocrates postulated in his book *Airs, Waters and Places* that the body is constructed of the same elements as the atmosphere and therefore the atmospheric disturbances that produce weather phenomena must also display in the body as symptoms of maladies. However, it was Aristotle, a half century later, who produced the first authoritative work on weather in his book *Meteorologica*, from which our modern term for the science is derived. The Greek word *meteor* is applied to anything high up, or reaching up – English equivalents varying from 'rising' to 'raised' and even 'noble'. *Meteorologica* strictly means the study of things on high, though Aristotle defines the study of meteorology as being confined to the area of space between the earth and the moon – the area in which comets and meteors appeared to exist. The heavenly bodies, he believed, moved in an element unknown on earth, an element he called 'ether'. The motion of the bodies within the ether was regular and could be defined and measured. However, below the ether came the four natural elements – earth, fire (or heat), water and air – whose movements, while originating from the influence of the eternal motion of the bodies in space, obeyed rules of their own. Earth, he reasoned, was the lowest of the elements, naturally residing beneath our feet, and any object made mainly of 'earth' fell through all the other elements back to the ground, its home. Water was next, since it descended as rain through the air, whereas air rose as bubbles in water. Fire was the highest of the elements, because fire alone was drawn through the air up to the heavens.

He dismissed the idea that winds were simply air in motion, as Hippocrates had suggested, for from where would the motion be driven? Instead he watched how rivers were formed from springs emanating from the ground and theorised that winds were the same, that the earth gave off airs that gathered together in currents and flowed towards some ultimate goal. He identified these airs, or breaths, as being of two different kinds, which he called 'exhalations'. When the sun shone on the hot, dry areas of land, the earth gave out a hot, dry exhalation, which, being hot, naturally ascended up into the fiery regions high in the sky. But when the sun shone over the sea, it forced out a cool, moist breath that did not want to rise as far. Instead, it formed clouds, which in turn brought rain so that the moisture could return to its watery region on the earth. Neither exhalation could exist without the other, though sometimes there was very little cold mixed with the hot and sometimes moisture predominated. The particles of the elements caught up in these exhalations would battle with one another in their struggle to extract themselves and return to their natural layer, hence causing the wind. It is an idea that has not totally left us, since we still go out to 'face the elements' on a wild and windy day.

He might not have got it right, but Aristotle's deductive reasoning is astounding. In an age where it was accepted that anything as powerful as a gale, as destructive as a flood, or as benign as the gentle rain of spring could only be attributed to supernatural forces, or gods, his questioning of established religious teaching must have been as revolutionary as Charles Darwin's overturning of the biblical story of the creation. Without knowledge of the laws of physics, without instruments to measure and mathematics to analyse, he could only observe; and having observed, he attempted explanations to fit his observations. He looked at the Milky Way, shooting stars and comets, before coming to earth to examine the sea and the winds, rivers and the earth. He examined the sky and clouds, looked at earthquakes and the way the wind blows. His wide-ranging theories took in storms and calms, heat and cold, thunder, lightning and rainbows. He suggested that rain is the condensation of water vapour and that hail, frost and dew are formed from water vapour, too, depending on the amount of moisture and heat present in the air. So all-encompassing were Aristotle's theories that they held sway over meteorological thinking for nearly 2,000 years. It was a pity therefore that, while a few of his hypotheses did come close to the truth, most did not.

When Aristotle strove to find philosophical explanations for the atmosphere surrounding him, he was not seeking to foretell its future behaviour. Yet the vagaries of the weather can dictate the difference between starvation and plenty, and it can hold the balance between life and death on the seas. The farmers, the sailors, the builders, the gardeners and everyone whose livelihood depended on it, all needed to know when, not why, the weather would change. So while Aristotle watched, pondered and speculated, practical men were busy with their own applications of meteorology.

As early as the fifth century BC, evidence can be found of columns in public places being used as 'parapegmata'. These were a kind of pegboard into which pegs were inserted, usually to mark wind directions, which were essential to navigation. The most spectacular of Greek weather stations, though, was undoubtedly the magnificent Tower of the Winds in Athens. Built by Andronikos of Kyrrhos in about 40 BC, this eight-sided stone tower served a dual purpose, one of which was as a public clock. Each side carried a sundial and, with eight sides, there was always at least one that was in the sun, so the tower could catch each of the hours between sunrise and sunset. At night a simple water clock would indicate the passing of time. Undoubtedly, though, its most unique function was as a primitive weather predictor. Each of its sides faced a different wind. Each wind was given a name and its special characteristics were represented on the tower by a carved face and a sculpture. If you stood with your back to the wind and faced the tower, it would tell you about that particular wind and the kind of weather you could anticipate while it blew. So, if you could get to Athens and place yourself in front of the Tower of the Winds, you might just be given some inkling as to what the weather in the next hour or so might bring. Of course, people wanted a more readily available and longer prognostication than that, and there were those willing to provide it.

One of these, Theophrastus of Eresus, produced a book on winds and weather signs, a work eminently more useful to the populace in general than was Aristotle's monumental treatise. Theophrastus was no mere rustic chronicler, though. He had himself written a *Meteorologica*, which in essence agreed with much of Aristotle's theories, although he had his own rather graphic explanation of an earthquake, which he compared to wind trapped in the bowels of the earth – flatulence being another interpretation of that handy Greek word *meteor*. However, his *De signis tempestates*, or *Book of Signs*, was a much more homely affair. It contained upwards of 200 maxims of weather lore, many of which were old established traditions, passed down through the generations. Some are based on the behaviour of birds and animals:

> An ass shaking its ear indicates a storm.
> If birds who do not live on water wash themselves,
>   it indicates rain.

Some are based purely on observation of past weather patterns:

> Whenever there is fog, there is little or no rain.
> If there is much rain in winter, the spring is generally dry.

And some reflect maxims we still quote today, such as a red sky at sunrise indicating rain.

Much as they would have liked to believe it, the good citizens of Athens realised from their own experience that weather lore is not infallible. Storms do not materialise every time an ass shakes an ear, and wet springs can follow equally wet winters. There were surer methods. If, as the Greeks did, you believed that the winds and the rain were controlled by the gods, then rather than attempting to predict the weather, you could ask for the gods' help in providing the weather you wanted. Offerings and sacrifices were made, and if you addressed the right deity, you could pray for rain to water your crops, calm for your harvest or a good wind and fair if you were setting out to sea. On the other hand you could, if you were privileged, consult the oracles to predict the weather for you. Another alternative was astrology. This was an age when astrology was accorded a standing the equal of the noble study of astronomy. Weather behaviour was recorded and analysed in conjunction with the movements of the planets, and just as climatic occurrences were associated with different deities, so they were attributed to the various aspects and conjunctions of planets within the constellations. The Greek astrologers would turn to the stars to predict the weather prospects for the coming season, in much the same way as we tune into a long-range forecast today.

After Aristotle and Theophrastus, there would be no significant meteorological developments to astound the world for the best part of two millennia. The Romans absorbed into their culture the wisdom of the Greek philosophers and seers, accepting Greek weather lore wholesale and adding little of their own. Virgil, in his first book of the *Georgics*, advises farmers on how to anticipate the changing of the weather, using the same signs that Theophrastus had spelt out to the Greeks. It is doubtful that Virgil came upon such detailed observations quite independently – it is much more likely that Theophrastus simply proved to be as reliable as anything the Romans could invent. Those public-spirited Greek citizens plugging pegs into columns to record the weather found their successors in Roman pontiffs, charged with the duty of noting on the calendar any occurrences of exceptional weather – storm, drought, heatwave, and so on. However, the importance of the date against which the phenomenon was recorded came not in the fact that it was, say, the Ides of March, but in the positions of the heavenly bodies at the time of the event. The next time that same combination of stars and planets was due in the sky, the presumed wisdom was that exactly the same weather would occur. Thus the stars, rather than the gods, became cast as arbiters of the weather for the coming year. Astrologers would draw up 'weather calendars', which the citizens fell upon as eagerly as they would any almanac of star-based predictions of events for the coming year. Such was the regard accorded to these weather calendars, they eventually began to appear in the *fasti*, the official town registers, alongside the days prescribed for legal and public business, the names of the magistrates and other public records. Yet, despite the 'officialese' of astronomical forecasts, their foundation was as shaky as the weather lore, based equally on an observed relationship between a particular weather sequence and an event – in this case a certain conjunction of planets rather than the crop of berries on a tree – a presumed correlation that, if it did indeed prove correct, came about more by happen-chance than scientific deduction.

In fact, the application of true scientific reasoning in a modern sense would be a long time in coming. The theories of the ancient philosophers dominated educated thinking until well into the seventeenth century, when, thanks to Galileo with his telescope and Isaac Newton with his theory of gravity, men started to throw off the blinkers of a classical education and began to observe, experiment and question the established knowledge. Even into the eighteenth century, some scholars were still reluctant to abandon the Aristotelian school, and the battle raged between the classical thinkers who supported the wisdom of the ancients and the challenge from the new breed of scientists who rejected those teachings.

Meteorology's ascendance from the dark ages of astrological hocus-pocus into the light of new science began with the turn of the seventeenth century and the invention of the thermometer. In Aristotelian theory, heat was a physical entity, one of the four elements, the one he called 'fire', and it occupied a place in space somewhere above the other elements: earth, water and air. The influence of motion from outer space caused the elements to mix, the presence of fire creating 'hot', and its absence 'cold'. Degrees of hot and cold had always been recognised as being subject to individual sensibility – something the Roman bather knew well as he progressed from the chilly *frigidarium*, through the sweating *caldarium* to the hot bath and finally into the cold plunge. How to measure temperature objectively, though, was a puzzle that had tested the inventiveness of educated men for centuries. The first inkling of a possible solution came in 1575, when the work of a first-century philosopher from Alexandria was published in Europe.

His name was Hero, or Heron, and his true provenance is lost in the mists of history. Who was Hero? Was he one man or several? When did he live, was it BC or AD? Which of his writings did he write himself, and was he writing of his own inventions or reporting on others? If everything attributed to him was actually his own work, the man was a genius. Some of his inventions were way ahead of their time, and had they all been developed, then the Alexandrians would have been buying their drinks from slot machines, watching mechanical puppet shows and powering their industry with steam. One of his books, the *Pneumatica*, details various mechanical devices operated by air, water and steam pressure, and it introduced the revolutionary concept of the elasticity of air to the Western world. The revelation that air expanded when heated was seized upon as a method of measuring temperature and the race began to capture the formless element of air and harness it into a workable thermometer.

Several men vied for the glory of being first. Cornelis Drebbel, Dutch-born inventor at the Court of James I, and Cambridge physician Robert Fludd independently produced similar instruments, while in the Italian academic hotbed of Venice, Santorio and Sagredo worked on projects of their own. But it is Galileo Galilei who is usually credited by history as being the inventor, around the end of the sixteenth century, of the thermometer, or thermoscope as it was then known.

Galileo's first instrument bore little resemblance to the decorative coloured balls bobbing up and down in the glass tube that are today sold as a 'Galileo thermometer'. These thermometers do, in fact, use a principle discovered by Galileo, that the density of water varies with its temperature – the higher the temperature, the lower the density. But his original thermoscope consisted of a long, thin glass tube, at one end of which was a bulb about the size of an egg. Galileo would cup the bulb in his hands to warm the air, then he would immerse the open end of the tube in a canister of water (so that it was upside down to the way we are accustomed to seeing the bulb and tube of a modern thermometer). When he removed his hands from the bulb, the air cooled and shrank, and the water was drawn up the tube to fill the vacuum.

The device worked well in that, with the bulb heated to different temperatures, the column of water drawn up the tube grew longer or shorter. But without a standard scale to measure against, it was of no practical use. At what temperature did the water boil? At what temperature did it freeze? And how did one measure temperatures above and below these points with a water thermometer? It would be another hundred years or so before Daniel Gabriel Fahrenheit invented the mercury thermometer, enabling him, Reaumur and Celsius to develop their universally accepted scales, and even longer until the Celsius scale was adopted as the scientific standard for temperature measurement.

Earlier in his life, Galileo had been fascinated by the Venetian galleys he liked to watch sailing into port and unloading their rich cargoes. He delighted in applying his mind to mechanical devices to improve the ships' efficiency, and one of these inventions was a suction pump. Galileo subscribed to the commonly held supposition that nature abhors a vacuum, and that was the reason why water would flow against gravity up into the vacuum created by suction. He noted, though, that there was a finite height to which the pump could raise water, and that this limit was around 33ft (10m). A quirk of nature such as this begged an explanation, but it was not to be Galileo who set about trying to find it. Evangelista Torricelli, his former pupil and successor as mathematician to the Grand Duke of Tuscany at his Court in Florence, was the one to take up the challenge.

Torricelli reasoned that it had to be the weight of the air outside the tube that supported the water within. Above 33ft, the weight of the water in the pump became greater than the weight of the supporting air and that was why Galileo's suction pump would lift the water no higher. That air had weight was another of those innovative ideas that the Alexandrian mastermind, Hero, had presented to the Western world. But people were sceptical and Torricelli's supposition that it was the air pressing on the surface of the reservoir of water that pushed the water into the vacuum, not the water being pulled by the attraction of the vacuum itself, needed some proving. Tubes of water 33ft high were far too unwieldy to work with, so in 1643 he directed his assistant Vincenzo Viviani to conduct an experiment with a denser medium.

He calculated that a 33ft column of water could be represented by a much smaller 30in column of 'quicksilver' (mercury). Using a tube a little longer than this, with a bulb at one end, he filled it with mercury, heated it to remove all the air, then had Viviani close the other end of the tube with his finger and invert it in an open vessel filled with mercury. When he released his finger under the surface, mercury ran from the tube until just over 29in of mercury remained supported above the surface in the vessel, leaving a vacuum at the top. Torricelli had invented the first barometer.

A rush of discoveries were to follow. Frenchman Blaise Pascal conceived the idea that the weight of air must be heaviest at sea level and should decrease with altitude. To prove this, in 1647 he persuaded his brother-in-law, Florin Périer, to climb the Puy de Dome mountain in Clermont, clasping a portable version of Torricelli's mercury barometer. Every few feet, the brave chap would stop to catch his breath and measure the height of the mercury. When he returned safely to the bottom of the mountain, Pascal made off with his list of readings and constructed a rough-and-ready scale of pressure-to-height ratios. The world now had a much simpler method of measuring altitude than the complex combination of sightings and trigonometric calculations.

It was Robert Boyle who brought the barometer to England and gave it its name – from the Greek *baros* meaning weight and *metron* as measure. He embarked on the series of experiments with gasses and vacuums that would establish, in 1662, the law that bears his name. Boyle's Law is concerned with the relationship between volume, temperature and pressure in a gas, and states that, given constant temperature and a constant amount of gas, if the volume of the container increases (i.e. the gas has more room and is therefore less dense), then the pressure of the gas decreases by a relative amount. Or, to put it another way, as the density of the gas decreases, so does the pressure, in equal proportions – gas at twice its original pressure occupies half its original volume. Therefore, as Périer climbed his mountain and the air (a gas) became less dense, the pressure of the air also became less. Boyle thought that a possible explanation was that a gas is composed of minute particles that, when pressurised, were pushed together into a reduced space.

So the barometer's use as a scientific instrument was developing, but from where came the impetus to propel it towards becoming the weather glass, the predictor of the weather? It did not take Torricelli long to realise that the height of the mercury supported by air pressure differed slightly each time he repeated his experiment. He figured that ambient temperature must have an effect, but, even at a constant temperature and altitude, variations occurred. From there it was not a big step to consider other obvious states of the atmosphere and their significance. Was the pressure higher when the air was dry rather than when it was wet? What reaction did the wind cause, stirring up the air and moving it at different rates? And what about sunlight and cloud? To record and compare observations of these various atmospheric phenomena, instruments were needed to measure them.

A basic hygrometer (to measure moisture in the atmosphere) had been in use since the fifteenth century. It was well known that certain materials absorbed water; water, or moisture, present in the air could therefore be detected and measured. Around 1450, Nicholas of Cues made a hygrometer from wool, attaching it to a scale to weigh the water absorbed, while Leone Battista Alberti chose a sponge for his hygrometer – an idea also used by Leonardo da Vinci, who made a drawing of the instrument and is therefore popularly regarded as its inventor. The principle of water absorption worked, but it was difficult to measure the minute changes of humidity that the developing science of meteorology required. A material that reacted to water in a more subtle way was needed.

Venetian Santorio Santorre experimented with the use of salts, cord and flax. Torricelli, wanting to compare the pressure of air at different humidities, looked around in the fields and alighted on wild oats. The awns or 'beards' of seeds react to moisture in the air by twisting, and various inventors sought to harness the property for an accurate hygrometer. It was to be an Englishman, Robert Hooke, who built the first successful seed-awn hygrometer.

Hooke's major claim to fame is for his *Micrographia*, published in 1665, containing drawings of his observations under a microscope, but he was also an astronomer, a surveyor (along with Wren, he worked on the reconstruction of London after the Great Fire) and a multitalented inventor who designed the forerunner of the respirator, the balance spring and escapement mechanism for clocks, and the first universal joint (a principle still used in cars today). Very much a hands-on scientist, he was eminently suited to the role of Curator of Experiments for the Royal Society, a post in which he employed all his quick-thinking inventiveness to produce new, practical experiments at the rate of one a week, to amuse and inform the eminent philosophic members. It did not leave him much time to devote to in-depth study of his own, yet he did find the energy to indulge his interest in the relationship of air pressure to weather, and to tinker with improvements to current meteorological instruments. To the barometer, he added a wheel that multiplied the movement of the mercury and recorded it with a rotating needle against a scale on a dial, enabling much smaller changes in pressure to be detected. And to measure wind speed, he developed his anemometer (from *anemos*, Greek for 'wind'), consisting of a disc held vertically in the wind, the strength of the wind being measured by the angle of incline it induced.

It was to be Isaac Newton who initiated the next step forward, pitching in with his laws of motion, the first two of which would prove crucial in the study of the movement of the atmosphere and the forces at work in it. Newton's first law states that the velocity of an object does not change unless a force acts on it, and the second that the force applied to an object accelerates that object at a rate that varies according to its mass. However, at this stage in the development of embryonic meteorology, the applications of these two laws were not understood. It is, though, his third law that is the ultimate stumbling block when it comes to taming meteorology into a logical science. That every action has an equal and opposite reaction is the very reason why the weather, with its myriad of actions and reactions, exhibits a behaviour pattern so difficult to measure, model and predict. Just how vastly difficult was, perhaps fortunately, way beyond the conception of the early pioneers of meteorological science.

Still, within the space of seventy years, science had provided all the vital instruments needed to begin the transformation of meteorology from a guessing game to a measurable science – and as a by-product, scientists had grasped the equally vital notion that air was a gas with weight, volume and pressure. The ability to measure was only a beginning. To interpret the readings and transform them into an understanding of climate needed more: it needed scientists to throw off the shroud of Aristotelian theory and discover a viable replacement. It would take a whole new mindset, new ideas, new methods and a transformation from individual effort to international cooperation on an unheard-of scale. But this would be many generations in coming. Man would first have to take his cue from Aristotle; to stand back, look and learn, in the light of new science, what the earth itself had to tell. Patient observation filled the gap over the next 100 years while meteorology marked time, waiting to begin a new era.

TWO

Weather Watchers

There had been a whole week of earthquakes, tremors that frightened the cattle and had the people casting a wary eye on the mountain craters as they went about their work. In Iceland, they were used to volcanic activity, yet this time they could feel the ground beneath them preparing for a great upheaval. On 8 June it began. The ground split first at the Laki fissure, some 200km west of the capital, Reykjavik. Volcanic explosions, one after the other, spread along 30km of the fault, spewing red hot lava, dust, ash and toxic particles into the air. For eight months the eruptions continued, and lava, preceded by a rush of meltwater, spread slowly into a basalt lake, up to 25km wide. Dust billowed into the atmosphere, ton after ton of it, and travelled wide across the skies of the northern hemisphere, trailing effects on the weather that were quite beyond living experience. Suddenly, the whole of Europe was a weather observatory; the topic was on everyone's tongue, and anyone who kept a diary must have noted its effects.

In the town of Selbourne, in Hampshire, Gilbert White wrote:

> The summer of 1783 was an amazing and portentous one, and full of horrible phenomena; for besides the alarming meteors and thunder-storms that affrighted and distressed the different counties of this kingdom, the peculiar haze or smokey fog, that prevailed for many weeks in this island and in every part of Europe, and even beyond its limits, was a most extraordinary appearance, unlike anything known within the memory of man . . . The sun, at noon, looked as black as a clouded moon, and shed a rust-coloured ferruginous light on the ground . . . but was particularly lurid and blood-coloured at rising and setting.

At Burford, in Oxfordshire, these strange phenomena captured the imagination of a 10-year-old schoolboy, dragging him away from his books and out onto the hills to watch the swirling dry mists and the huge red disc of the setting sun. His name was Luke Howard and the experience was to give birth to an absorbing interest in the sky, especially the clouds, their size and shape, and the weather they brought in their wake. In fact, Luke Howard's enduring contribution to meteorology would be to classify and give names to the various types, or 'modifications' as he called them, of clouds.

Born in 1772, Luke Howard was a product of an eighteenth-century grammar-school education, which pumped into boys, with the threat of the cane, the classics, most specifically Latin, and excluded anything vaguely approaching science. 'My pretensions as a man of science are consequently but slender,' wrote Howard; 'being born, however, with observant faculties, I began even here to make use of them, as well as I could without a guide'. Evenings and weekends would find him gazing at the sky from the Cotswold ridges that surrounded his school in Burford; or closeted in his room performing experiments with whatever substance he could lay his hands on; or taking readings from his barometer, making notes of its measurements and concurrent observations of the weather outside his window. 'The numerous Aurorae boreales in those years interested me. I settled in my mind one remarkable configuration of the Clouds in a full sky because it was of rare occurrence.'

After school, Luke was apprenticed to a retail chemist in Stockport, Cheshire, a far more receptive outlet for his scientific leanings. Unrestained by the specialisation thought necessary today, he studied French and botany alongside chemistry, and in 1796 he partnered a man called William Allen. Allen concentrated on the chemistry business while Howard took over the running of their laboratory in Plaistow, London, where botany became his main field of work. His boyhood fascination with meteorology, though, was not forgotten. He says 'in passing between the works and my dwelling, I resumed the observations I had long been making on the face of the sky, and began to keep a Meteorological Register'. By 1806 he had established this register into a daily pattern, recording not just atmospheric pressure, but maximum and minimum pressure, maximum and minimum temperature, wind direction, rainfall and evaporation. A year later, the editor of the *Athenaeum* began a column he called the 'Meteorological Register', which appeared every month as a record of Luke Howard's observations from his laboratory in Plaistow.

Howard's mastery of and dedication to the barometer was eventually to lead, in 1847, to the publication of his work *Barometrographia*, which at the time, with its curves, graphs and diagrams showing the variation of barometric readings over the years and how they change with the weather, was to be the definitive work on the subject. However, although this was a fine achievement, it is not his work with the barometer that is Luke Howard's major contribution to meteorological science; it is his boyhood fascination with the sky. Those early notebooks he kept by his bedroom window, the notebooks of barometer and thermometer readings and weather aspects, would have been full of descriptions of the clouds he saw – grey, white, translucent, heaped, layered or wispy like strands of curly hair. When, as a young man in his laboratory in Plaistow, he began again to record his weather readings, he realised the need for a formalised system of noting cloud forms. Until then, no one had attempted to classify clouds. They were looked on as too random and too fleeting for recognisable forms to be identified. However, in his observations Luke had discovered that clouds could be divided into classes and types. All he needed was a formal means of naming and describing them, and it was to his botanical background that he turned for inspiration.

Howard was a talented botanist as well as a chemist and in 1800 he read a paper on pollens to the Linnean Society, a society that took its name from the influential eighteenth-century Swedish botanist, Carl von Linne. Von Linne, better known by the Latinised version of his name, Carolus Linneaus, created a system of botanical classification that was organised hierarchically, specific elements within generic groups, thus giving plants the Latin names familiar to botanists and gardeners today.

Using this as a basis, Howard invented his own system for naming the clouds. Those schoolday hours spent over Latin exercises were to come into their own in the names he chose for the three distinctive classes he identified:

- *cirrus* – filaments, or strands, which stretch or curl like hair, growing in all directions;
- *cumulus* – clouds that form conical heaps, piling upwards from their base; and
- *stratus* – wide, extensive, horizontal sheets of cloud.

These basic classifications could be used in combination – *cirrocumulus* and *cumulostratus*, for example – and with the addition of *nimbus* to identify a rain-bearing cloud, they described briefly and precisely any cloud formation. Clouds, as Luke Howard says himself, 'are commonly as good visible indications of the operation of these causes [causes that affect all atmospheric variations] as is the countenance of the state of a person's mind or body'. In 1804 he presented his ideas in an essay he read to the Askesian Society.

> My friend Allen and myself belonged also to a select Philosophical Society, which met every fortnight during the winter (at his house) in London; each member being required by the Rules to bring in an Essay, in turn for discussion, or pay a fine. It was the obligation thus contracted which occasioned me to present to that society . . . the 'Essay on Clouds'.

In the essay, he explained his purpose in defining the clouds, stating that:

> in order to enable the Meteorologist to apply the key of Analysis to the experience of others, as well as to record his own with brevity and precision, it may perhaps be allowable to introduce a methodical nomenclature applicable to the various forms of suspended water, or, in other words, to the Modifications of Cloud.

And so good was his system, it forms the basis of the cloud names still in use today. Certainly, at the time, it was a huge step forward, since the burgeoning army of amateur weather observers could now record their cloud observations in a uniform manner, allowing direct comparison, place to place, and analysis of the weather, temperature and barometric pressure that the clouds brought with them.

Although Luke Howard is credited with being the first man to name the clouds, the first published attempt at classification actually came from a French naturalist in 1802. Jean-Baptiste Antoine de Monnet, the self-styled Chevalier de Lamarck, was a talented scientist who had the insight to conjecture some of the most remarkable ideas of his time – and the misfortune to be overshadowed by others on each and every one of them. Although he worked totally independently of Howard, four out of Lamarck's five types of clouds coincide with Howard's basic classifications. However, Lamarck's work attracted little attention, even in his native France – Luke Howard's use of the Latin he had despised as a child probably accounted in part for his system's ready international acceptability.

Lamarck's study of the clouds was far from being his first foray into the world of meteorology. Around 1780 he got together with fellow French scientist Antoine Lavoisier – the man renowned for discovering oxygen – to thrash out plans for predicting the weather. Lavoisier noted that

> the facts needed for this art [of weather prediction] are: regular, daily observation of variations in the height of the mercury in the barometer, the force and direction of the wind at different elevations, and the humidity of the air. With this data it is almost always possible to predict one or two days in advance, with a high probability, the weather which will occur; I think, too, that it would not be impossible to publish every morning a prediction which would be of great use for society.

Lavoisier and Lamarck did not just talk about their ideas, they began to put them into action. They distributed a number of instruments to various locations in France, and planned to extend their observation stations across Europe, even throughout the world. Their plans might well have come to fruition had not Lavoisier lost his head under the blade of a Revolutionary guillotine in 1794. Lamarck attempted to continue their work, and in 1800 began to publish regular meteorological reports. However, like Howard, he was first and foremost a biologist and was employed as such by the Emperor Napoleon. Napoleon took exception to his biologist spending most of his time predicting the weather and ordered him to stick to natural history. You did not argue with Napoleon. So by one means or another, the world's first attempted weather-prediction service became a victim of the French Revolution. The idea was not yet ripe for revival.

Lamarck, though, did launch forth with yet another of his big ideas, this one derived directly from his role as a biologist – a theory of evolution that preceded Darwin by half a century. The Lamarckian view was known as the 'inheritance of acquired characteristics', and differed from Darwin in that he suggested it was the result of a creature's own efforts – for example, a giraffe's neck is long as a result of generations of giraffes stretching up for the topmost leaves – that was passed on to the next generation. Along with Charles Darwin's own grandfather, Erasmus, Lamarck was one of the first evolutionists, whose views were rejected in their time, only to resurge some forty years later, to be reconsidered by the new radical thinkers, such as Darwin.

Howard, Lavoisier, Lamarck, all with their detailed observations and weather diaries, were just a few of the many, all over Europe, who were keeping a watch on the weather. Would-be meteorologists everywhere would rush out each morning with their thermometer and barometer, their weather-eye cocked to the sky. These people might be serious scientists, but more likely they would be the retired mariner with time on his hands, or the country rector taking a break from his flock. One of the earliest amateur weather diaries still in existence was produced by just such a gentleman, the Revd William Merle, vicar of Driby in Lincolnshire. The Revd Merle took his daily observations assiduously from 1337 until 1344, partly at Driby and partly in Oxford, and his diary, called *Consideraciones temperiei*, is preserved at Merton College, Oxford. No doubt there were others, now lost, throughout the Middle Ages, continuing the tradition established by the philosophers and astrologers of ancient times.

Possibly the first to attempt an organised meteorological register was Robert Hooke. Hooke, who had improved the barometer by adding a dial, sought, by observation, to turn the instrument into a predictor of weather, the 'weather glass'. If he could get enough people to record daily the state of the weather at various heights of the barometer, he reasoned that he could determine the highest probability for each reading and add it to his dial. In 1663 he contrived a 'Method for Making a History of the Weather' and handed out to a select number of volunteers instructions for taking daily observations. A specimen table of Hooke's scheme of weather observations was printed just six years later in *History of the Royal Society* by Thomas Spratt. Interestingly, under the column marked 'The Quarters of the Wind and its Strength', the strength of the wind is shown represented by numbers, much like the Beaufort scale – except this is 240 years before Beaufort defined his scale. Disillusioned by the lack of enthusiasm, he gave up. However, he must have acquired some readings, enough to produce those rough-and-ready weather guides we still see inscribed around the dial of a home barometer – 'Change' at 29.5in, with 'Fair' 'Set Fair' and 'Very Dry' for progressively higher readings and 'Rain', 'Much Rain' and 'Stormy' for the lower.

There were two different purposes in keeping a weather diary. One was to record the weather at a fixed time every day in order to accumulate sufficient readings to search for weather patterns, as Hooke was attempting to do. However, without a purpose, continuous daily observations were onerous and uninteresting. The second purpose, favoured by the majority of the early weather-watchers, was to record, not the tiny nuances of change in ordinary weather, but the extraordinary, the inexplicable, the spectacular eruptions of the elements. They were meteorologists in the classical sense, in that what they sought were 'meteors' and an explanation for them. These would indeed include comets (which were still thought to be terrestrial in origin), but the term covered anything dramatic – giant hailstones, sky-splitting lightning flashes, a deluge of rain, strange lights in the heavens. Any appearance would have them putting pen to paper, describing, even drawing, what they saw. It was an exciting race for recognition – to be the first to predict when the next comet would appear, which combination of circumstances tempted out the Auroras, or what force generated the power to strobe the sky with blinding flashes and roll the thunder. And they were taken seriously. The *Philosophical Transactions of the Royal Society* throughout most of the eighteenth century are filled with multifarious reports of meteors.

One man who bucked the trend was a Cornish clergyman, William Borlase. Evidently, his parochial flock were a contented and troublefree bunch, so leaving their vicar with time on his hands to indulge in a detailed study of his Cornish home. Equipped with a quadrant, a theodolite, a microscope and hydrostatic balance, several thermometers, barometers, hydrometers, numerous pens, copious sheets of paper and a bottomless well of inquisitiveness, the intrepid reverend set out to chronicle Cornwall. The results, the *Natural History of Cornwall*, appeared in print in 1758. The title sheet lists its contents, which embrace everything he saw: the rivers and the coast, rocks and ores, flowers and trees, the birds, fish, insects and animals, and even the activities of its human inhabitants, their language and customs, arts and trades. Plus, of course, the climate and the weather.

Cornwall, jutting out into the Atlantic on England's south-west tip and cut off from the world by the River Tamar, might, in those days, have been a country as foreign as Mongolia to the population of the English metropolis. The Revd Borlase's book was a revelation, not only in the breadth of its information about this hitherto unknown corner of the kingdom, but in the depth of its scientific content and the skills of its writer. Dr Thomas Hornsby, Savilian Professor of Astronomy at Oxford, braved the 300 miles and several days of rutted cart tracks in order to venture beyond Plymouth into the unknown, and meet this extraordinary Cornishman. He persuaded Borlase not only to continue with the daily weather diary he had begun for his book but also to extend its breadth, and from 1768, his weather summaries appeared regularly in both the Royal Society's *Transactions* and the *London Chronicle*. Borlase would continue with his daily observations until his death in 1772, yet even he was unconvinced of their usefulness. He wrote to a friend in London, saying:

This may in time either facilitate some more perfect theories of winds or weather in our climate, or, which is altogether as likely, show the uncertainty and vanity of all such attempts; in short, the atmosphere is such an irritable mixture . . . that nothing permanent and sure is to be expected.

William Borlase might well have died, and his observations with him, yet his sponsor, Dr Hornsby, was to ensure that the discipline of daily recording he had started would continue, almost without interruption, from then until the present day.

In Woodstock Road, Oxford, in the garden of Green College, is an octagonal tower, over 30m high, bearing a distinct resemblance to the old Tower of the Winds in Athens. Not quite as old – this tower dates from 1772, the year of Revd William Borlase's death – but the shape is significant, since this tower, too, has a meteorological pedigree. It was built for Dr Thomas Hornsby, and its purpose was as an astronomical observatory. Equipped with the newest of astronomical instruments, the Radcliffe Observatory was set to provide a magnificent facility for Hornsby and his students. However, there was one major problem for an astronomer attempting to take detailed measurements of the heavens from a land-based telescope, and that was the atmosphere. The air distorts the image reaching the eye – as water appears to distort the dimensions of fish in a tank – and an allowance for refraction needs to be factored into the equations. As with anything to do with the atmosphere, the rate of refraction is not a constant – its effect varies with humidity, pressure and, most significantly, temperature. So all observatories had to make meteorological measurements as a preamble to the important task of observing the stars.

At most observatories, these measurements went unrecorded for posterity, but Dr Hornsby, no doubt inspired by his encounter with Borlase, began in 1767 jotting down the observations he took and keeping them carefully alongside his astronomical readings. His behaviour was apparently regarded as idiosyncratic by his colleagues, since, after his death in 1810, the records ceased. They were not to be taken seriously again until 1849, by which time, thanks to Luke Howard and those who were to follow him, meteorology had turned a significant corner. Astronomical observatories all over the world had become the hotbeds of meteorological progress, and Oxford's Radcliffe came to join them. The old instruments, originally installed for Hornsby, were removed and were replaced by newer, better and more accurate ones. Radcliffe's observations were published again, beginning in 1849 in the annual reports, then daily from 1853. Today, the Radcliffe Astronomical Observatory will not be found in the garden of Green College. It has moved to Pretoria, thousands of miles away from Oxford's polluted atmosphere. But the Meteorological Station it gave birth to still exists, on its original site, and still returns its observations as a small but important cog in a worldwide network of weather stations. The station's status may now be insignificant, but the Radcliffe records hold an importance to climatology second to none – for they are one of the oldest sets of complete, continuous meteorological records in existence.

By the end of the eighteenth century, science, besides being a serious subject of study, was being adopted by the moneyed classes as a social attribute. Geology and natural history were all the rage; 'collections' were becoming as essential an item of drawing-room furniture as the chairs themselves, despite the fact that few collectors knew anything about the objects they collected and even fewer took to the field in pursuit of their hobby.

In the first decade of the nineteenth century, the barometer, or weather glass, was fast becoming the latest fashionable 'must have' in the homes of the rich and noble. The 'stick' type of barometer was common at that era; the glass-fronted wooden case would be ornately carved and the metal scale plates etched in a flowery, cursive script, the whole instrument a luxurious and highly prized work of art. The instrument stood about a metre high, consisting of a glass tube closed at its top end standing in a small cistern of mercury. The pressure of the atmosphere forcing down upon the surface of the mercury in the cistern held a column of mercury in place in the glass tube; the higher the pressure, the higher the column it would support. The height was measured against a scale at the side of the column.

Despite the weather glass's imposing and ornamental housing, as a predictor of the weather its ability was poor. In the hands of an enthusiast, it could provide an indication of what the elements might produce in the imminent future, although even these predictions relied for their accuracy entirely on the currently existing knowledge of the relationships between atmospheric pressure and weather behaviour. The amateur, observing solely the legend written beside the barometric scale, would find his expectations frequently confounded. Even this limited insight, though, was available only to those who had access to a barometer, and in 1810 these were very few people indeed.

Fortunately, there is a readily available method of divining the same indications, and it is free to all. Sitting permanently above, the sky can reveal much. If clouds are a uniform grey, the sun will not appear for some time; if they are light and fluffy in a blue sky, it will remain fine; if they begin to pile up, rain will follow soon; and if they tower, black and threatening, storms are imminent. Those who worked among the elements and whose livelihoods depended upon them had, over the centuries, developed a much keener sense of the coming weather than the aristocrat could glean from the barometer sitting in his hallway. Luke Howard was beginning to give a scientific basis to the association of clouds to future weather, yet these people – the sailors, gardeners, farmers and shepherds – were already reading the skies, plus the many other signs in nature, to predict both short- and long-range weather forecasts. Indeed, it is to a shepherd that the most popular book of weather lore was accredited, the anonymous Shepherd of Banbury.

Whether a particular Shepherd of Banbury ever existed is doubtful, but within this compact little book were distilled essential rules of weatherwise wisdom that had obviously been accumulated over many years. There is no mention within its pages of a barometer, or air pressure, or temperature (other than warm or cold), yet this book catered for a need no scientist could yet fill – the need to be forewarned of the weather. As Theophrastus, Aratus and Virgil had done in classical times, the shepherd eschewed theory in favour of observation. Yet these are no mere ramblings, for they show evidence of careful observation, possibly an accumulation of weather diaries long since lost. The Shepherd of Banbury was popular. His book, *The Shepherd of Banbury's Rules*, appeared in 1670, and there were at least three more editions during the eighteenth century – and probably any farmer who could read had a set of the Shepherd's rules tucked away on the shelf, along with the Bible. John Claridge's foreword to the 1744 edition states:

> Every thing, in time, becomes to him a sort of weather-gage. The sun, the moon, the stars, the clouds, the winds, the mists, the trees, the flowers, the herbs, and almost every animal with which he is acquainted, all these become, to such a person, instruments of real knowledge.

Of the sun, the Shepherd says: '*If the sun rise red and fiery –* wind and rain'; and of the clouds, for example, he says: '*If cloudy and the clouds soon decrease –* certain fair weather'; of mists: '*If Mists rise to the Hill-tops –* Rain in a Day or Two'; or of rain: '*Sudden Rains never* last long: *but when the Air grows thick by degrees and the Sun Moon and Stars shine dimmer and dimmer,* then it is like to rain six Hours usually.' The Shepherd is not afraid of long-range forecasting, either: '*Observe that in eight Years' Time there is as much South-West Wind as North-East, and consequently as many wet Years as dry.*'

Yet even the Shepherd was not infallible, and towards the end of the eighteenth century people were becoming disillusioned. They longed for accurate prediction, but theorists could offer no hope – despite moves in science, they were still groping along with Aristotle and finding little sense. As for those mountains of carefully collected data on meteors, they defied analysis. There were no patterns to be defined, meteors abided by no discernible law, so that scientists were forced to agree with the religious writers when they declared that man cannot fathom God's work, nor trace His causes. When it was proved around 1780 that comets had extraterrestrial origins, even the Royal Society abandoned its collection of meteoric observations. Meteorology seemed destined to wander in the same random wilderness as the weather, with no official scientific backing.

However, meteorology had simply been biding its time. As the old century faded into the new, a North of England scientist by the name of John Dalton embarked on a series of observations and revelations that would begin to transform all the old conceptions of the weather. A contemporary of Luke Howard, Dalton was likewise a typical late-eighteenth-century philospher who could turn his hand and his agile brain to many branches of science. Principally a chemist and physicist, he is best known for his work on atomic theory. He was, though, no mean weather-watcher, and with a collection of over 200,000 observations taken in his lifetime, Dalton could ensure his place in a meteorological Hall of Fame on that achievement alone. But his contribution did not stop there. His abiding interest was hydrology – the study of rainfall and water circulation within the atmosphere – no doubt inspired by the frequent soakings he would have endured as a native of the Lake District, the rainiest corner of England. Even in his day, the hydrological cycle of the earth was not properly understood – some still clung to the centuries-old ideas of springs emanating from mountain-top reservoirs fed from the sea by a mysterious underground complex of waterways. Dalton swept these antiquated theories aside when he proved that precipitation alone was sufficient to feed the rivers and springs. He went on, too, to develop further the work of Robert Boyle, some hundred years before, and made two vital discoveries. The first was that as air was compressed, its temperature rose and, vice versa, as air expanded, the temperature dropped – meaning that air could create its own heat, independent of the sun. And the second was that besides the gasses that make up air, the atmosphere nearly always contains water particles, particles so small that they cannot be seen, and that warm air can hold more of these invisible particles than can cold.

Man was finally on his way to a new understanding of the forces at work in the atmosphere, and of the nature of the furious power that unleashes a storm.

THREE

Storm about Storms

One day in 1821 a storm tore across the North American state of Connecticut, uprooting trees and leaving vast swathes of forest flattened in its wake. While the population gathered up the debris, one man set out into the shattered forests, scrambling over felled trees, measuring and drawing the positions in which they lay. He noticed that the gale had not cut a straight swathe through the trees, for the fallen trunks in various parts of the forest were lying in different directions – in fact, trees 40 miles apart had been lain flat by winds blowing in totally opposite directions. It seemed incredible to him that winds of such ferocity could blow in completely opposite directions at the same time within such a small distance. What was more, from the lie of the trees, it was apparent that within the storm the winds had whirled around a central point in a counter-clockwise direction and that this central point had taken a curved path across the forest.

William C. Redfield was a Connecticut man, a saddler by trade. He cultivated a preoccupation with storms that he was to invest in a thoroughly practical manner as his interests swung from saddles to steamships and he set up a steam navigation line between New York and Albany. The early steamboat was not the most comfortable or safe way to sail. The dirty, smelly monster that was a primitive steam engine churned away in the bowels of the ship, threatening to explode at any moment, and a voyage was made even more intolerable when the entire vessel pitched and rolled in a storm. By 1821 sailors knew a lot about storms and how to recognise their approach. A sudden drop in the barometer, a build-up of mounting cumulus clouds, an ominous change in the wind, all heralded a storm and put the wary sailor on his guard. What was not known, though, was why storms developed, how they progressed and what their behaviour was as they passed. Redfield determined to find out. He wanted to discover whether the pattern he had observed in the Connecticut forest was merely a freak of nature or whether it was normal behaviour for winds always to blow in a circular fashion around the centre of low pressure and for the direction of flow always to be anticlockwise. For ten years, as his ships chugged up and down the eastern seaboard of America, he measured and plotted the readings he needed to test his case and by 1831 he was confident that his findings were correct. He launched into print, stating that the winds in a storm circled the centre in an anticlockwise whirlwind, except in the southern hemisphere, where the wind mysteriously changed around and blew in a clockwise direction. From these conclusions, he went on to theorise that it must be the centrifugal force created by the swift, circulating winds that caused the air pressure to reduce at the centre of the storm.

Unfortunately for Redfield, he had lumbered into a field already occupied by learned, scientific men, completely overturning their ideas – and they were not willing to have them overturned. Back in 1783, the University of Mannheim had conducted its own meteorological study, much in the way that Lavoisier and Lamarck had done in France, issuing instruments to various observers and coordinating their readings. These detailed, controlled observations from all over Germany were published in an almanac known as the *Mannheim Ephemerides*. In 1826 a German professor of mathematics and physics by the name of Heinrich Brandes came upon this unique collection of statistics and constructed a series of weather charts, from which he purported to show that winds blow from the perimeter of a depression directly towards the low pressure area at the centre. Redfield's new claim, that, rather than heading to the centre, winds in a storm whirled in circles around it, came as a complete contradiction to Brandes's theory.

Brandes himself was not drawn into direct conflict with Redfield. That was left to Redfield's fellow American, a man named James Pollard Espy. Espy had studied law, then changed his interest to classical languages, and it was while he was earning his living as a teacher of languages that he had come across the theories of Brandes. Captivated by the whole concept of storms and storm theory, he took Brandes's charts, studied them and added his own claim that, at the centre of a depression, the winds combined into a column of air that rose and flowed out over the top of the storm. Rising air, it was known, gave a low pressure reading on the barometer – the faster the rise, the lower the pressure. What was more, Espy suggested that the rising air released a heat, or energy, that fuelled the storm and led to the growth of clouds and thence to rain. He launched an acrimonious attack on Redfield, who, he argued, offered no scientific explanation for his centrifugal theory. Redfield countered that his evidence for the circular direction of the winds was soundly supported by observation, and, although he could offer no reasons for the phenomenon, the fact that it existed was beyond question. Rather than taking Redfield's findings and adapting his proposals in the light of the additional evidence, Espy refused to budge – what he said must be true, he held; his theory proved it. The argument became public, took to the printed page and developed into a slanging match between the two, in which not just theories but personalities were slated. American scientists began to take sides. Prominent among Espy's supporters was Joseph Henry, Professor of Natural Philosophy at Princeton.

Tall, blond and handsome, Henry had, as a young man, been fascinated by the theatre and yearned for fame and fortune as an actor. In a period of inactivity following an illness, he alighted upon a book of scientific experiments, which so mesmerised him that he suddenly changed his allegiance from theatre to laboratory and embarked on a career as a scientist and schoolteacher. Enthralled by the magic of electricity, Henry loved to tinker. Through his experiments, he discovered the properties of the electric relay and the electromagnet and delighted in demonstrating his own designs of electric motors and primitive telegraph in front of an audience of schoolboys in his classroom. For Henry, the thrill was in the invention, not in the writing of it, and as a mere schoolmaster, he was reluctant to exhibit his work in the critical arena of world-class science. Over in England, Michael Faraday had no such hang-ups, and when the Englishman's similar inventions were published and patented, Henry immediately regretted his own diffidence. Faraday's subsequent celebrity was to rankle with him for the rest of his life, reawakening the former would-be actor's desire for fame and spurring him on with a determination to ensure he kept to the winning side in the future.

In his work with electricity, Henry had undertaken investigations into storms. Espy's theories bore a fundamental correlation to his own and he was excited by the possibilities opened up by them. With the Redfield/Espy battle projecting the pursuit of storm theory into the public eye, Henry sighted future glory for the eventual victor and decided to pitch himself in on Espy's side. The scientific community as a whole was veering gradually to Espy, too, closing ranks around him and leaving Redfield, the untrained but practical observer, cut off on the outside. A weaker man might have given way, but Redfield was a determined fighter, albeit a fighter in need of an ally – which he was to find in the unlikely person of an English Army officer, William Reid.

A lieutenant-colonel of the Royal Engineers stationed in Barbados, Reid first arrived on the island after it had been devastated by a hurricane in 1832, charged with supervising the reconstruction of government buildings there. As he stepped off the ship, he could not believe his eyes. It was as if the island had been struck by an earthquake – boats were shattered and washed up on rocks, roofs had lifted, walls had tumbled and entire buildings lay in ruins. It was like nothing he had witnessed back home in England and he was curious to discover more about the power unleashed in a storm, a force that could tear apart solid buildings with the sheer strength of its blast. Some of Redfield's papers had been published in Britain, and Reid had read them before he left. He wrote to William Redfield.

The letter could not have been better timed nor more welcome; a correspondence began that developed into full collaboration. Like Redfield, Reid was a practical man. While Redfield was motivated by a desire to aid mariners in surviving storms at sea, Lieutenant-Colonel Reid, from his base in Barbados, sought to provide some means of forewarning an island population so they could look to their own safety before a coming storm.

Between them, the two men spent several years investigating and observing storms. While Redfield was occupied in the Western Atlantic, Reid's various British Army postings allowed him to observe the hurricanes south of the equator, in the Indian Ocean off Mauritius and the Reunion

Islands. The results led Redfield to conclude that storms and hurricanes originated somewhere between the equator and the tropic. North of the equator, he found, the storms travelled northwards until they reached 30° latitude, the northern extent of the Trade Winds, then, on encountering the prevailing westerlies, they were deflected to the east and crossed the Atlantic towards Europe.

Meanwhile, the feud in North America between William Redfield and his opponent, James Espy, continued to deepen. It overflowed into publicly aired arguments concerning tides – perceived atmospheric tides in which cold and warm air ebbs and flows over the surface of the earth. Redfield, in his steamship voyages, had observed that in certain parts of the Pacific Ocean sea tides were not governed by the pull of lunar gravitation. He surmised that in that vast expanse of water a centrifugal force generated by the rotation of the earth overrode the moon's influence, and suggested that the same force might well be exerted on the mass of atmosphere above the earth. Espy quashed this argument, claiming that atmospheric tides were generated not by the moon but by the daily heating of the sun. While Redfield, unsure of his ground on this one, allowed Espy his point, he refused to buckle under and join the 'in' crowd of American meteorologists now accepting Espy's 'calorific', or heating, mechanism in the creation of storms. In 1835 nature provided them all with another opportunity to demonstrate their theories on the ground.

The tornado, or 'Brunswick Spout' as it became known, crossed New Brunswick, New Jersey, in 1835, leaving a landscape littered with fallen trees. Scientists flocked to survey the devastation, each keen to read the evidence for himself. William Redfield came, as did James Espy, with Joseph Henry and others following in his trail. To the casual onlooker, trees appeared to have toppled in confusion, yet not so to men who knew what they were expecting to find. A triumphant Espy looked at the chaos and declared that the storm as it had passed had drawn the gales to its centre, a perfect proof of his theory. Redfield, who visited on at least two occasions to look at the selfsame trees, saw a completely different pattern, evidence of winds whirling around a central point, just as he had first distinguished in the fallen trees of Connecticut fourteen years previously. The Brunswick Spout, therefore, could be shown to collaborate both theories and the intellectual storm raged on.

One notable effect of the controversy was that it drew into itself some of the best scientific brains in America, who fuelled it and kept it on the boil in much the same way as the physical storms they were studying. Joseph Henry was conspicuous at the forefront, as was his friend, Benjamin Franklin's great-grandson Alexander Dallas Bache, and a young Elias Loomis, who would later become a world-renowned expert on cyclones and the way in which they travelled. Such interest ensured that the study of storms progressed at a far greater rate than it would perhaps otherwise have done. Thus far, though, the Atlantic had proved an effective barrier to this particular furore and the scientific communities of Europe had remained immune. It was not to last, as the various proponents prepared to take their theories across the ocean and embark on a series of European lecture tours in order to win the Europeans to their particular point of view.

Lieutenant-Colonel Reid watched as his partner in research became more bitterly embroiled in the arguments, but apart from warning Redfield to leave it be – which warning Redfield chose to ignore – he tried his best to keep his distance. However, it was Reid who first introduced Redfield's ideas to the British when in 1838 he travelled home to England to present his paper on the Law of Storms at the Annual Meeting of the British Association for the Advancement of Science. The British Association had been founded a few years earlier in 1831, as a forum for scientists to meet together to discuss and promote different aspects of science. Unlike the more venerable Royal Society, which insisted that prospective members should have achieved some scientific merit, the British Association offered membership to anyone, and its annual meetings were an opportunity for scientists to air their ideas and research for comment, without having to comply with any strict rules. A new president was elected each year, who would hold the annual meeting in a major town of his choice, a tradition that has remained unbroken, apart from in the years of the two twentieth-century world wars, until the present day. In 1838 it was the turn of the Duke of Northumberland to host the conference, and the venue was Newcastle-on-Tyne.

Reid read to the assembly a workmanlike paper, in which he supported the theories of Redfield. His ideas, published later that year in his first book, *An Attempt to Develop the Law of Storms*, were practical in nature, giving straightforward guidance to sailors on how to steer out of the worst difficulties in a storm. His rules were based on Redfield's theory of the circulating nature of the wind around the centre of the storm. Reid's idea was to show a sailing ship captain how to manœuvre his ship away from the direction in which the storm was moving, rather than following the same course as the storm and becoming trapped in it. However, it was not Reid's words, but the comments on them from eminent British scientist Sir John Herschel that were to flame the Redfield/Espy controversy on the eastern side of the Atlantic.

Sir John was the son of the famous astronomer Sir William Herschel. He had followed his father into astronomy rather than take up his own interest in chemistry, partly because he wanted to complete the work his father had begun, mapping the heavens. He also had his father to thank for the understanding of meteorology that he had acquired from the work they had done together in compiling tables of lunar observations in combination with aspects of the weather. Sir John had come to Newcastle fresh from five years at the Cape of Good Hope working on a survey of the southern skies. No doubt the two-month sail back from the Cape had provided him with first-hand experience of Lieutenant-Colonel Reid's subject matter, though his professional interest would stem from the sunspots he was then studying – another project that had originally been his father's – and his theory that 'spots', or dark patches, on the sun were caused by tornado-like forces ripping holes in the luminous material covering the sun's surface. Sir John, with the example of his father to follow, was a meticulous observer, recording everything first then theorising only when he had the proof before him. His well-meant advice to Reid was that, although his observation was good, he really ought to advance a theory of his own, one that would please the scientists and enable them to remember the arguments he had submitted. He went on to applaud Redfield and his work, but he derided Espy and his calorific heat theories.

The news soon filtered back to Espy in America. Criticism from such a respected scientist as Sir John Herschel could not be ignored, even at a distance of 3,000 miles. But Espy did his best to shrug it off, putting it down to Sir John's incomplete knowledge of his work, since not all his papers would have been available to him in England. However, it made Espy doubly determined to travel to Europe. If he could eliminate the apparent lack of understanding, he felt sure he could win over the European scientific community – men like Sir John Herschel, John Dalton, Michael Faraday, George Airy and John Daniell in Britain, and Frenchmen François Arago and Louis Joseph Gay-Lussac – in the same way as he had the American. He had another motive, too. The French Academy was offering a substantial prize for a theory to explain the movement of the atmosphere, and Espy was convinced he could win.

In 1840 James Espy arrived in England to begin his tour, and he was invited to Scotland to present a paper at the British Association's annual meeting, held that year in Glasgow under the auspices of the Marquis of Breadalbane. However, his hopes of overawing the members of the British Association with the brilliance of his ideas were to be dashed. He had left it too late. Lieutenant-Colonel Reid had already won over the majority to Redfield's point of view, and they were not convinced that the complex forces at work in a tornado could be adequately deduced from a few fallen trees. The Glasgow experience was repeated everywhere he went. Reid summed up the British reaction to Redfield: 'I hear from England that people's minds were satisfied with the revolving theory of storms [i.e. Redfield's]; so that few cared to listen at Glasgow and elsewhere to Mr Espy's explanations of his particular theory.'

Redfield himself did not make the journey to Europe. Lieutenant-Colonel Reid, as his ambassador, had represented his theories perfectly well. However, Reid had not ventured beyond Britain's shores with his talks and his writing, which left the rest of the European continent much more open to the ideas of Redfield's rivals. Espy was keen to exploit this. He left the cool of London for the warmth of Paris, where he was greeted with an exuberant Gallic reception. The loyal American press delighted in reporting his triumph. Unfortunately, Espy missed out on the French Academy's prize for a theory of the atmosphere, since the committee decided his theory was not general enough, but they were certainly impressed by Espy's work. Indeed they went so far as to recommend to the American government that Espy should be supported by official funding. This was a somewhat two-faced move on the part of the French, considering that they refused to back his work with their own finance, yet nevertheless the recommendation must have carried weight with the US government, since, in 1842, Espy was assigned to the office of the Surgeon-General of the US Army, where his duties were to prepare a daily weather map, and where all the resources of the Army Medical Corps were available to him to help continue his investigations.

After this European exposure, the feud between Espy and Redfield began to die down, a spent force. A couple of strong, stubborn characters, they had managed to keep the row running for around twenty years, splitting the American scientific community in the process. Had they been able to work together, the cause of meteorology would have been much better served, for the future was show that both men had valid ideas to offer. Redfield was sound in his observations and more sophisticated simultaneous readings would soon support his proposition concerning the direction of the winds. However, he would have been taken more seriously had he not ventured into scientific realms he did not fully understand, since his attempt at a theory, that centrifugal forces cause low pressure at the centre of a storm, was wrong. It was Espy who was nearer the mark here: his suggestion that it was generated heat, or a latent caloric release, that energised the storm was in essence a significant step forward in the laws of thermodynamics, but his stubborn refusal to accept the evidence of practical observations of the direction of the winds robbed him of the recognition that was rightfully his.

Lieutenant-Colonel Reid, while remaining constant in his support of Redfield, refused to let the dispute with Espy deflect him from his original goal. He had not forgotten the sight of the hurricane damage that had so appalled him when he first arrived in Barbados. The island was a frequent target for such destructive attacks of nature, and Reid figured that if the islanders had some sort of warning of the approach of these storms, they would be able to do something to help themselves – battening their buildings, for example, and making sure their boats did not set out to sea. In order to make such a warning, though, he needed to recognise the telltale signs of a hurricane's approach, and this meant understanding the nature of them. The collaboration with Redfield had provided much of this understanding, but he needed more. Leaving the bickering to the Americans, Reid had been quietly and systematically following his own schedule of observations and drawing his own conclusions. As an officer of the Royal Engineers, he was fortunate enough to travel extensively, but his postings were not designed to position him necessarily in the exact location he would wish in order to further his meteorological studies. By 1838 he was realising the limitations of attempting to obtain meaningful observations single-handedly. Needing to cover a gaping hole in his data for the Bay of Bengal and the Arabian Sea, he cast around for likely assistance and it was this need that had brought him to Britain with his paper on storms to present before the British Association. While well applauded, it did not produce any practical offers of help. Thinking laterally, he alighted on the notion of approaching the East India Company, whose trading posts covered the exact area of the world he wished to investigate. Official cooperation was not forthcoming, but his letter did eventually land on the desk of one Henry Piddington, whose personal interest was aroused by Reid's request.

Piddington offered his services and set about retrieving the observations Reid needed, producing his first report in 1839. Piddington, though, did not end his input there. Stimulated by Reid's dedication, he continued his research for a further fifteen years, until in 1855 he distilled the sum of his knowledge into the strangely named *Sailor's Horn-book*. In the book, whose full title went on to describe itself as *The Law of Storms in all Parts of the World*, Piddington developed in much more detail Reid's earlier *Attempt*, and taking his rules for sailors, he both extended and simplified them, suggesting to sailors how they could take a cut-out circular shape of a storm marked with its circulating winds and move it across a chart to produce a practical model of how the wind's direction would vary through the storm. It was in this book that he described how the winds did not just circulate around a depression, but curved in a spiral towards the centre, and he coined the term 'cyclone' from the Greek word for 'coil of a snake' to describe the effect. Thus, it was not a scientist, but an employee of the British East India Company who finally sewed together the theories of Redfield (circulating winds) and of Espy (winds drawn towards the centre of the storm) – and he reached his conclusion without the help of a single fallen tree.

Meanwhile, William Reid monitored the progress of his own and Piddington's researches until he felt confident that the path and nature of a storm could be fairly well predicted by the changes in air pressure, and he campaigned vigorously for a storm-warning system for the island of Barbados – a campaign that bore fruit in 1847 when a storm signal station was set up at Carlisle Bay, based on a self-recording barometer.

Despite this success, Reid realised that his warning system was severely limited. What could work for one island ought to be available for the rest of the world, but that meant understanding all types of storms and storm behaviour over all the continents and oceans. Even with Henry Piddington's sterling efforts, he had barely made a start on the mountains of observations needed. What he desperately wanted was official funding and an extensive workforce to provide the geographical breadth of readings required. He decided to attempt to breach the stone wall of the Civil Service.

It is not everyone who can walk into the Colonial Office with a direct request for assistance, plus the people and funds it would require, and be successful, but Lieutenant-Colonel Reid managed it – no doubt much aided by his second book, published in 1849, *The Progress of the Development of the Law of Storms*, which had triggered scientific interest in his work. Armed with official sanction to initiate meteorological observations in the colonies, Reid approached his former commanding officer in the Royal Engineers, Major-General Sir John Burgoyne. Reid had hit upon just the right man. Burgoyne, now inspector-general of fortifications, commanded an extensive staff at various onshore locations throughout the colonies, and what is more, he gave an enthusiastic reception to Reid's purpose, recognising its worth to an army scattered in storm-threatened positions throughout the world. He authorised the setting-up of a chain of meteorological stations at his various colonial outposts, strategically placed to make observations across a wide swathe of the globe. The officer he appointed as overseer of the observatories was another Royal Engineers' officer, Captain Henry James. James would prove to be an inspired choice. Now in his late forties, he had spent much of his life observing and making measurements on land as both engineer and surveyor. He had recently completed a tour of duty in Ireland on the Ordnance Survey Triangulation of Britain, a major undertaking that was still some years from fruition. He set to work on this new assignment with assiduity and within months the stations were providing more information than Reid had managed to accumulate in years of solitary work.

Reid, while delighted with Henry James's flood of statistics, desperately needed to extend observations into the unpenetrated territories of the United States. He discussed the problem with Burgoyne. Ever supportive, Burgoyne used his influence to initiate an approach to the US government through diplomatic channels. His request for cooperation, along with the instruction booklet that had been issued to the Royal Engineers' observers, was sent by the British Legation in Washington to Mr Daniel Webster, US Department of State, on 13 November 1851. Webster, who professed no knowledge of meteorological matters, nor much of an idea of who was in the best position to answer, made two copies of Burgoyne's letter and passed them on.

One copy came via the Secretary of State for War into the hands of Joseph Henry, one of the combatants who had backed Espy in the storm over storms. It was an appropriate choice. Joseph Henry, in his position as the first chief of the Smithsonian Institution (a post obtained in 1846 through the influence of his close friend Alexander Dallas Bache), had established a network of over 150 volunteer weather observers across the American Midwest. The 1,500 square miles of open plains were an ideal location in which to study the behaviour of storms, and from the monthly returns he received from the field, Henry could collect and analyse the incoming data in his offices in Washington. The results from those observations taken in conjunction with the Royal Engineers' own readings would extend world coverage in exactly the way Burgoyne and Reid had envisaged.

Unwittingly, though, Daniel Webster had thrown Burgoyne's letter slap bang into the middle of yet another American scientific tug of war, this one over the extent and control of meteorological observations. Joseph Henry, again, was one of the major protagonists, supported by Bache. His opponent, and the man who received the second copy of Sir John Burgoyne's request, was a naval lieutenant, Lt Matthew Maury, Superintendent of the National Observatory. On the face of it, he would seem a strange choice – a naval officer in charge of an astronomical observatory – but Maury was a man with a mission, and his vision involved mapping the winds over the world. For ten years now, he had been forced to make do with the collection of meteorological data he could obtain from American maritime sources, Joseph Henry having blocked access to his land observers. The British had what Maury could only dream of – a series of outposts under their control, scattered about the globe in various different climates, and an army, literally, of men available to be ordered to make all the observations that could possibly be required. He determined to take advantage of Burgoyne's offer, with or without the cooperation of Joseph Henry.

Charting the Winds

If he had not fallen from a tree at the age of 12, Matthew Fontaine Maury would not have been sitting, in 1851, in Foggy Bottom, near the Potomac river in Washington, in charge of the USA's prestigious National Observatory. In fact, before he climbed that tree one dusty morning in 1818, the only future he could see for himself was a life of toil as a poor farmer in backwoods Tennessee. From 45ft up in the treetop, the view in all directions would have been of rough, pioneer country – fields hacked out of the woods and the ramshackle shacks that were his farmhouse home. A trail, little more than a path, snaked through the trees towards the small crossroads village of Franklin some 3 miles away, where Matthew received what little education he had been given at the small field school. Why Matthew had climbed the tree is not clear – probably for the same reason that any boy climbs a tree; because it was there. But what is certain is that he came back down the quickest way possible, tumbling all 45ft to the ground, where he lay unconscious.

Matthew's father, Richard Maury, had brought his young family to Tennessee in search of a better living seven years earlier, when Matthew was just 5. They came from Spotsylvania County, near Fredericksburg, Virginia, where Richard's farm had failed, reducing the family to near destitution. He loaded his wagon with their belongings and drove for weeks across dusty prairies, up the 100 miles of mountainous Wilderness Road and through the Cumberland Gap, while young Matthew, his mother, brothers and sisters trudged alongside. Richard was the twelfth son of the prolific Revd James Maury of Fredericksburg, a descendant of the Maury family who had fled to America from France as persecuted Huguenots at the end of the seventeenth century. Matthew's mother, Diana Minor Maury, was of Dutch descent, emanating from a sea captain by the name of Maindort Doodes, who had sailed to America, liked it and settled in Virginia in 1671.

Doodes's seafaring blood manifested itself first in John, Matthew's eldest brother and ten years his senior. He succeeded in persuading his father to let him join the Navy – a big sacrifice for Richard, who must have relied on the help of a strong young man in the running of the farm. Left behind to cope with the chores, Matthew relived his brother's colourful life through the long, exciting letters that arrived frequently from all parts of the world. Farm life did not suit him. He longed to go to sea, too, but knew this was impossible. His father needed him at home.

When they carried him inside, still unconscious from his fall, it seemed unlikely he would live. He did survive, though his back was so severely injured that the doctor declared it would be years before he would be fit for physical work – if indeed he ever would. Matthew's best chance of a future, therefore, would be with an education, so his father scraped together the money and sent him to the nearest college, Harpeth Academy. It was quickly evident that Matthew was academically brilliant, soon outstripping what the small college could offer. It was also evident that his powers of recovery were equally outstanding, since he rapidly grew fit and strong. It was a double-edged blessing. On the one hand, he would be physically able to undertake his dearest ambition, a life at sea, yet on the other, he was once again indispensable on the farm. It was brother John who set the seal. Having made his family proud by rising rapidly to the rank of Flag-Captain of the Fleet, he died in service, aged 28. The early death of their eldest son devastated his parents. When Matthew proposed following his brother into the Navy, Richard adamantly refused. The Maurys would not risk the life of another son. In a fit of desperation, Matthew did something he had never even contemplated before – he defied his father, going behind his back with a secret application. He could not hide, though, the official letter that came in return, offering him a position as midshipman in the US Navy. In the resulting row, his father denied him any assistance, so Matthew borrowed a scraggy mare and set off, penniless, on the 750-mile ride to Washington.

Traditionally, 'midshipman' was the first rung on the ladder to becoming an officer, and a typical midshipman would be the son of a moneyed and well-connected family, taken on as a boy, usually at 13 or 14. At 19, impoverished and of farming stock, Matthew Maury was far from typical. Distancing himself from the drunken rowdiness of the midshipmen's mess, he used his time at sea in self-education. He devoured the ship's books, such as they were, on navigation, mathematics and astronomy, and he practised spherical trigonometry by chalking out problems on the ship's round-shot ammunition. He sailed twice to Europe, rounded Cape Horn, spent time stationed in Valparaiso, South America, where the predictability of the tropical winds fascinated him, and nearly lost his life when his ship, the *Vincennes*, was taken by a strong current off Nuku Hiva, the largest of the Marquesas Islands in the South Pacific, and just missed being dashed helplessly against a cliff.

After passing his midshipman's examination at the age of 25, he became engaged to his cousin Ann Herndon, then was posted as sailing-master on the *Falmouth*, on which he once again sailed to South America via Cape Horn. One of his duties was to maintain the ship's logbook, noting the course steered, the distance travelled, the direction of the winds and their speeds, and the ship's position obtained by astronomical observations. These made good, valuable data, but Maury was baffled by what happened to them when they were returned to port at the end of the voyage. Why did the Navy simply bury them in their stores? Why were they not available to other seafarers? Why did they have to navigate relying on out-of-date textbooks and sketchy or non-existent charts, when on every voyage they assiduously recorded new and better information? Maury decided to keep his own logs in addition to the ship's logs, and from them he published his first paper, a set of sailing directions for the navigation of the notoriously dangerous coasts of Cape Horn. Promotion to lieutenant rapidly followed and with it his marriage. Posted to the steamship *Engineer*, he surveyed the south-east coast of the USA, taking soundings of sea depths and currents off North Carolina, Georgia and the Gulf of Mexico. His abilities were enormous and his rapid rise through the ranks would have been assured had it not been for an overstaffed navy and a lack of naval vessels. It was on a further spell of enforced shore leave that Maury's life was set to change tack once again – occasioned by yet another fall.

Matthew had gone to visit his mother and father, failing in health yet still living on their primitive farm in Tennessee. His objective was to persuade them to move to his own home in Fredericksburg, to be near the countless Maury relations. While he was at Franklin, he received an official letter, forwarded on by his wife Ann, with orders to report to his next ship, the *Consort*, due to sail shortly from New York. The half pay of shore leave was difficult to live on, and with sea postings few and far between, he had no option but to go. The letter had already taken several days to reach him, and time was short, so Maury bid farewell to his parents and booked himself onto the next stagecoach out. On 17 October 1839 the coach came to the staging post in Lancaster, Ohio. Here there were three more passengers waiting to board the already overloaded vehicle, and Maury gallantly gave up his inside seat to squeeze on top. The road was stony and winding, and as it plunged into a particularly sharp corner, the inevitable happened. Top heavy with extra passengers and excess baggage, the coach toppled over. Along with the cases, bags and boxes, Maury shot from his perch and hit the ground hard, his leg bent beneath him. With a fractured thigh and a dislocated knee, he spent the next seven weeks in agony, lying in a shabby inn, unable to move. The *Consort* sailed without him.

Maury limped back to Fredericksburg, worried his injury would not only ruin his naval career, but render him incapable of earning a living at all. However, after a couple of years languishing inactive on the Navy's lists, he was offered a shore posting in Washington, as Head of the Navy's Depot of Charts, a division of the Bureau of Ordnance and Hydrography. The role of the depot was to act as a store not just for charts, but for all the instruments – the sextants, quadrants, chronometers, thermometers, barometers, and so on – for issue to naval ships. Although the pay eased his financial straits, Maury found little challenge in acting as a glorified storekeeper, nor in the routine of cleaning and recalibrating the instruments.

However, it did give him the chance to investigate the riddle he had worried over on the *Falmouth* – what did happen to all those thousands of logs handed in by returning ships? The answer was nothing. They lay where they had been put, undisturbed, some of them, for decades. Delving into boxes, he grabbed handfuls of papers at random, and found himself looking at records of weather conditions, wind, temperatures of the sea and the air, depth soundings, currents, tides – a mine of invaluable information, left to rot. Logs existed for all seasons from all parts of the globe. Some were dashed and haphazard, some beautifully detailed, while others made him smile where a young officer, bored on watch, had doodled cartoons across the paper, or amused himself composing limericks. In that dusty storeroom of abandoned logs, Maury found his destiny: he determined that from this chaos he would create a series of charts that would present sailors with some return for their hours of painstaking record keeping, charts that would assist future captains to benefit from the experience of their predecessors. It was a daunting task, for there was no order to any of the mounds of paper, and it would take months, years even, of one man's effort to make a careful analysis of all these old logs and distil the results into a useful form.

He immediately applied for extra staff – twelve naval officers, plus a clerk and a draughtsman – and while he was waiting for them to materialise, he pondered on the best way in which the mass of data could be used. He decided to make a start on examining everything he could find on the New York to Rio de Janiero route, the route he had himself logged on board the *Falmouth*, and eventually his goal crystallised into the creation of a set of sailing instructions that would point to the fastest route for the journey. In a sailing ship, the shortest route is not necessarily the quickest; the ship that makes the best progress is the ship that sails by the most favourable winds. The trick would be to plot charts that showed where to find these winds, and how to avoid the areas in which storms or calms were most likely to occur.

The old logs were all very well but they followed no set form, and Maury soon found gaping holes in his data. To plug them, he needed fresh, orderly observations and he was fortunate in finding ready cooperation from the Navy, which allowed him to redesign the format of the logs and to hand them out to naval vessels along with the charts and instruments issued from his depot. So far so good, but, with only thirty-seven ships at sea, the Navy was going to take a long time to provide all the new information he required. He cast longing glances at the stream of mercantile vessels sailing in and out of the east coast ports, the traders and the coasters and those setting off for whaling trips way up into the Arctic

Circle. With the cooperation of the National Institute for the Advancement of Science, he published a general appeal for them to assist him in his charting.

> To enable it to bring this undertaking to a useful issue, the Bureau [of Ordnance and Hydrography] relies much on the public spirit and intelligence of American owners and masters of ships. It takes this opportunity of inviting their co-operation . . .

The invitation was refused. Mercantile shipowners were not ready to let some government busybody come spying into their operation, no matter how well intentioned he might sound. Maury's only option was to show by example, and he worked hard, sixteen hours a day, driving himself to the limit, until he produced, within the year, his first set of sailing directions. Still the merchants were not impressed – but his masters were. At last he received some of the staff he had requested, and his tiny Depot of Charts was crammed with bodies, desks and papers. Now he had the staff, he needed the space for them, and he cast envious eyes to the other side of town, where, at Foggy Bottom on a low hill alongside the marshy banks of Washington's Potomac river, the Navy was erecting a brand new building with a rotating copper-sheathed dome to house the Naval Observatory. Even when the astronomers and their equipment were installed, there would be ample room for Maury's expanded Depot of Charts. In 1844 his wish was granted when he won the post of superintendent of the observatory, which, with the latest in optical telescopes and an able staff, was also to become the nation's National Observatory. Maury moved himself, his desk and his reams of papers into the equatorial room, a spacious accommodation he shared with the 8.9in 'comet-seeker' glass.

His appointment had attracted stiff opposition. Alexander Dallas Bache, a crony of Joseph Henry, had put forward his own candidate for the post, and Bache was not accustomed to being refused. Having recently landed the prestigious post of Superintendent of the United States Coastal Survey, Bache was establishing himself as a force to be reckoned with in American scientific circles. Capable of exuding considerable charm when needed, he was gathering around him a following of sycophants, men who in return for their allegiance received Bache's backing and support. With his new posting, Maury had gained enemies – not just Bache, but the whole of Bache's confederacy. They were everywhere, his supporters, and one of them, Sears C. Walker, lurked among the astronomers of the observatory. Walker resented his new boss's lack of proven astronomical expertise. Maury was a self-taught but able astronomer, as any navigator of worth aboard a sailing ship had to be, and he was also an efficient Superintendent. But Walker seethed under his leadership and in 1847 Maury was forced to dismiss him. Bache immediately hired him, then sent him to the Smithsonian Institution, now headed, at Bache's instigation, by his top protégé, Joseph Henry. Walker took with him a detailed abstract of his work at the observatory and made it available to Henry. Passing it off as Smithsonian research, Henry published both locally in the *Transactions of the Smithsonian Institution* and in Europe, in a German journal. Maury was aghast. Henry was a scientist of standing, much admired by Maury, and the two had only recently exchanged a pledge of mutual cooperation in the field of meteorological observation. The deceit had to be entirely Walker's, hoped Maury, but Henry's oblique attempt at denial was difficult to believe and the seeds of mistrust were sown.

In the spring of the following year, Maury's work on his charts was to receive the fillip it needed. The *Baltimore American* published a report of the voyage of the *WHDC Wright* in which Captain Jackson of Baltimore had followed Maury's route to Rio and clipped 35 days off the normal 100 days for the return journey. News of savings like that circulated quickly and suddenly US merchant shippers everywhere were clamouring for Maury's charts. Maury exploited this demand with a promise – you provide me with more and better observational data, and I will provide your charts free of charge. Soon 5,000 copies had been issued, along with sets of logs for the mariners to complete. The amount of good, well-formated data rocketed. With all the information being returned, Maury could contemplate expanding his horizons. He began to consider transatlantic routes and decided to reformat his charts into 'pilot charts', or what eventually became known as wind charts.

The idea took root from some English charts drawn up in 1831 by Commander A.B. Becher of the Royal Navy, which Maury had come across in his early delving into the old depot stores. Commander Becher was one of the Admiralty's ablest sea captains and a foremost surveyor. He recognised just how valuable charts of wind directions would be in enabling a pilot to pick the most advantageous sailing routes, so on a tour of duty in the Indian Ocean he had assiduously taken weather observations over much of the ocean and at various times. He then divided his chart of the ocean into squares and within the squares he noted the prevailing winds at the different seasons. Using the charts, a sailor could then plot his course to increase his chances of catching the winds that would best speed his voyage. Unfortunately, as a one-man team, Becher had no hope of achieving what he set out to do, and work was sketchy and incomplete. The Royal Navy Hydrographer's Office had enough on its hands with conventional surveying, so did not have the resources to build upon Becher's ideas. Maury, however, seized upon their potential. He divided the whole of the North Atlantic into squares representing five degrees of latitude by five of longitude. Within each square, he plotted the various readings he had gathered for that area into a 'wind rose', giving the figures, as Becher had done, for the prevailing wind and weather for different seasons. By 1849 he had issued his first copies of his wind charts of the North Atlantic. They were phenomenally successful – yet he knew that with more readings they could be much improved. He needed international cooperation, especially from the British.

In 1848 he tried his cousin, author Sarah Mytton Maury, who lived in Liverpool, but she did not have the right contacts or influence. So he turned to the American Association for the Advancement of Science, hoping for its help. However, with Bache and Henry installed as leading members, his appeal was blocked. He must have thought his prayers had been answered that morning in 1851 when he sat at his desk and picked out of his pile of mail the copy of the letter from Major-General Burgoyne offering the resources of the British Royal Engineers.

𝒩ations Confer

You will please . . . give your views whether any useful cooperation, direct or indirect, could be furnished by our vessels at sea with the instruments usually furnished to them; or at any of our navy yards, either with their present instruments or by the aid of others to be furnished for that purpose; and, if so, at what yards such observations would be most desirable, having regard to the observations of this kind which are known to be made at different places in connection with the Smithsonian Institute and public observatories.

Such were the instructions to Maury from his superior officer at the Bureau of Ordnance, Commodore Morris, when he sent him Major-General Sir John Burgoyne's request for reciprocal weather data. Burgoyne's intention was simple: that America provide additional observations so that the scope of his, or rather William Reid's, current study of storms could be expanded. And Commodore Morris's plea to Maury was equally simple: can we or can we not comply, without involving the Navy in any additional effort or expense? As it stood, though, it came nowhere near advancing Maury's maritime wind-charting aims. For one thing, Burgoyne had provided a pamphlet of detailed instructions on the observations he wanted and the methods of obtaining them as compiled by the coordinator of the Royal Engineers field observers, Captain Henry James. They did not fit what Maury required and neither, he suspected, would the Smithsonian Institution nor any of the dozens of private observatories be much impressed with being forced to bend their procedures to British dictates. If the USA imposed the British system without amendment, he replied to Morris, 'it would create confusion among our observatories, and be as likely to retard as advance the progress of meteorological research in the United States'.

Secondly, Burgoyne's plan concentrated specifically on observations taken on land. While Maury was more than happy to embrace land observations into his scheme, Burgoyne (and Joseph Henry, too) would need to accept the necessity of including maritime data in the enterprise.

> Five-sevenths of the surface of our planet are covered with water. It will be perceived, therefore, that . . . investigating the laws which govern the general circulation of the atmosphere, we must look to the sea for the rule, to the land for exceptions. Therefore no general system of meteorological observations can be considered complete unless it embrace the sea as well as the land. The value of researches conducted at this office with regard to the meteorology of the sea, would be greatly enhanced by cooperation from the observations on the land.

And why restrict the operation to England and the USA? Why not invite other countries such as Russia and France, where meteorological studies were already well advanced, into a worldwide consortium of nations, each one providing the meteorological data from both its ships and its land observatories.

> The atmosphere envelopes the earth, and all nations are equally interested in the investigation of those laws by which it is governed . . .
>
> For these reasons, therefore, I respectfully suggest that as an amendment to the British proposition, a more general system be proposed; that England, France, and Russia be invited to cooperate with their ships, by causing them to keep an abstract log according to a form to be agreed upon, and that authority be given to confer with the most distinguished navigators and meteorologists, both at home and abroad, for the purpose of devising, adapting and establishing a universal system of meteorological observations for the sea as well as the land.

Morris must have been dumbfounded by the flood of ideas put forward by the superintendent of his observatory. He sent Maury's effusive reply trickling up the diplomatic chain back to Burgoyne, yet before the letter had had a chance to leave the USA, never mind arrive back in London, the US Navy Department had seized on one line in Maury's letter:

> I beg leave to suggest a meteorological conference.

This was not a direct reply to Burgoyne; this was a different proposal entirely: an international conference to establish a system of observations not just on land, but at sea as well. In view of its maritime implications, Maury was given permission, as an officer of the US Navy, to bypass Burgoyne and confer directly with the British in an attempt to arrange such a conference. While Maury was delighted with this development, he did have one very big obstacle to overcome before he could set in motion any conference encompassing both sea- and land-based elements – he needed to get the Smithsonian Institution on his side.

Immediately after the New Year, he composed a careful letter of reconciliation to Joseph Henry, suggesting that they sink their differences in order to serve the greater good of the science of meteorology. Henry, of course, already had his own copy of Burgoyne's request, to which he had offered the simple exchange Burgoyne required – the results of his own land observations of American Midwest storms for the output of Burgoyne's and James's Royal Engineers' data. He saw no necessity of replying to Maury's letter. Come March, tired of waiting, Maury called for a coach and drove along the river and up the Mall to the red turreted sandstone edifice that was the Smithsonian Institution. Even as he bumped over the rickety wooden bridge, under the budding branches of the trees and up to the entrance, he must have been bracing himself for the interview to come.

Twenty years on, Henry was still smarting from Faraday's stealing the acclaim for his electrical inventions and he was not about to allow himself to be sidelined again, least of all by a naval non-scientist. Gauging the strength of Henry's feeling, Maury made the concession of removing himself from any involvement in a land-based operation, restricting his own contribution to the maritime side of the proposed project. It appeared to appease Henry – that is, he did not openly express any opposition – but what Maury did not know was that, with the help of Bache and his cohorts, Henry was already working towards establishing an independent operation with the British.

With the American side sorted, Maury was now ready to confer with the British about a proposed conference, as he had been bidden, but he had no idea whom to confer with. Approaching Burgoyne direct did not seem like a good idea – Maury's replies on that score were currently percolating through other channels as far as he was aware – and Burgoyne, he guessed, would not be too receptive to a plan that completely outstripped his own. The Royal Society was eventually suggested, and it was there that Maury duly addressed his request, but in the muddle of diplomatic exchanges it was redirected and landed up anyway at the major-general's door.

Burgoyne was dismayed. Maury had barged in and usurped his lead in what was after all a scheme that had been initiated by his own Royal Engineers. With a military man's distrust of foreigners, he objected to the participation of non-English-speaking nations in the proposed conference.

> I would beg to submit . . . that . . . the first effort should be made only by the British and United States conjointly . . .
> . . . being only two, they would collect a more manageable body for deliberation; they would intercommunicate in the same language; and, it is believed, have the same weights and measures.

It did not help that, in its convoluted journey, Maury's proposal had had a note attached to the effect that it was believed the idea had originated from the Royal Society. No such thing, declared Burgoyne, well and truly ruffled; the entire plan had been drawn up by his officer, Captain James. He told Addington of the Foreign Office that, since he, Addington, thought Sabine and the Royal Society were dealing with the matter, then perhaps he should go off and refer the whole thing to them.

The referral duly went to the Royal Society, addressed to Lieutenant-Colonel Edward Sabine, himself an eminent figure in meteorological circles, an interest that stemmed from his contribution to the international effort to study the earth's magnetic fields. As in astronomy, magnetic observations and measurements vary according to the vagaries of the weather; therefore meteorological and magnetic measurements are inextricably intertwined. Terrestial magnetism dictated the compass point and was essential to navigation, yet before the nineteenth century the differing directions of magnetic north and astronomical north had presented a puzzle. By 1832 a theoretical solution had been put forward by a Swede, Carl Friedrich Gauss, but his law needed practical proof, and a concerted effort involving scientists and observatories all over the world had been launched to provide it. Britain was an enthusiastic contributor, and foremost among the participating British scientists had been Edward Sabine. He personally had been instrumental in setting up magnetic observatories in places as far flung as St Helena, Cape Town, Hobart, India and Canada – in which latter the Americans had a hand, via the Smithsonian Institution and its chief, Joseph Henry. Sabine and Henry, therefore, had already been collaborating before Matthew Maury came onto the scene.

Poor Maury, it appears, had been projected into a situation already seething with pre-existing alliances and rivalries – and the Sabine–Henry connection was only one of them. The aims of the Royal Society were entirely scientific, therefore science (or 'philosophy') was the crux of its concern. Apart from an acknowledgement that 'correct climatological knowledge has [an important bearing] on the welfare and material interests of people of every country', these academics did not really consider it their prime thrust to concern themselves unduly about such worldly matters as sailors fighting their way out of cyclones at sea, or merchants stuffing their money chests with an extra purse of gold. However, it had been bothering the learned gentlemen for some time that it was proving impossible to tie meteorology down as a true science. Unlike magnetism, the atmosphere refused to subject itself to any discernible law. There was something elusive about its definition, an element of hit and miss in measuring its properties. Meteorology as the science of the atmosphere had developed as a branch of physics, with an element of chemistry and more than a fair sprinkling of inspired guesswork based on haphazard, uncontrolled observations. It had none of the precision of other fields of science – astronomy, for example – although most of the leading observatories had attached to them, by necessity, a department of meteorology. These, and the new magnetic observatories that also monitored the weather, already formed a worldwide network of meteorological stations. What they lacked was a system of coordination.

A few years earlier, steps had been taken to rectify this, when in 1845 European meteorologists had met at a conference in Cambridge. The conference certainly stimulated interest in meteorological research, and delegates had gone away flushed with the intention of circulating the results of their findings to one another. It did not work, or at least it worked only spasmodically, since without a central body to coordinate them, abstracts of readings were circulated only when each observatory got around to it. And then there was the difficulty of comparing like with like, since everyone used different systems (temperature alone could be measured in scales of Fahrenheit or Celsius or Reaumur), the very problem identified by Sir John Burgoyne. The Royal Society also knew something else that Maury did not know, though he might have guessed it, since he recognised the same problem in the United States. It was that European scientific establishments, jealous of their own national systems, were none of them willing to subjugate themselves to the dictate of another's.

Given all this, and perhaps Sabine's connection with Joseph Henry, it was not surprising that, when Sabine produced his report on Maury's plan on behalf of the Royal Society in May 1852, the conclusion concerning land-based observations was:

> With reference to the proposal for the establishment of a uniform plan in respect to instruments and modes of observation, the President and Council are not of the opinion that any practical advantage is likely to be obtained by pressing such a proposition in the present state of meteorological science.

However, this did not mean Maury's entire proposition was rejected out of hand. The Royal Society loaded praise upon Maury's wind charts, which were proving quite remarkable in chopping days off average sailing times. This was evidently a very practical application for international cooperation among maritime nations. However, practical applications as opposed to scientific investigations were not the proper concern of the Royal Society. Consequently, Sabine had no qualms in suggesting another authority to undertake it. 'The proposition of Lieutenant Maury to give a greater extension and a more systematic direction to the meteorological observations to be made at sea, appears to be deserving of the most serious attention of the Board of Admiralty.'

With the onus thus placed squarely back into government hands, Sabine happily launched into detailed recommendations about the provision of instruments on ships and the establishment of a central office, adequately staffed, to receive the observations and publish charts and sailing directions. The government was not quite so happy. Who was to foot the bill for this central office and 'adequate staff'? Between May and December, the proposal and a mounting number of papers went the rounds of the various departments. The Board of Trade declared it was prepared to circulate to merchant ships any forms supplied by the Admiralty or the Royal Society. It agreed that the exercise would be useless without staff to examine, tabulate and publish the results, but it was not prepared to pay for such staffing. Neither were the Lords of the Admiralty. They said, if the government were serious in wanting this project to go ahead, then it was up to the Treasury to provide extra funds. But the Treasury's purse stayed shut. Coastguards, lighthouses and naval ships already made a host of weather observations, it pointed out, and it did not at this stage see the need for any more – although it did concede that it might sanction a well-considered plan on the subject at a future date. The Treasury's suggestion was that perhaps some representatives of Britain and the United States ought to get together and discuss the matter.

After a whole year and many man hours of effort, the plan had come full circle, back to Burgoyne's original proposal. By then, Burgoyne himself had withdrawn, sickened by Maury's usurping of his idea; Joseph Henry had successfully deflected Maury from his observational territory on land and was continuing with his own work; and Maury was busily engaged in independently canvassing a raft of foreign countries on the subject of his proposed international marine conference. Delighted with the Royal Society's recommendation to the Admiralty, he had – or he thought he had – the British government on board, and with that the likelihood of other nations following suit was reasonably assured. However, what he had not wished to happen was for the British to bluster in and take charge of a conference that he desperately wanted to be America-led, or more to the point, led by Maury himself. Arbitararily, he had set August 1852 as the date. Then, purposely planning to detract from the British and to draw in the French, he had proposed Paris as the venue and confidently contacted the attachés of all the great sailing nations of Europe, South America and Asia. However, if he had expected a quick result, then he was heading for a disappointment.

Unfortunately, the Paris of 1852 was not the best choice of European capital to host his conference, since France was in the throes of yet another of those constitutional crises that had beset it on a regular basis since the Revolution of 1789. Secondly, the diplomats to whom he addressed his communications would have been as flummoxed as the British had been – the purpose of the conference was obviously scientific, not a concept the average public servant was equipped to deal with. August 1852 had come and gone with no conference in view.

Come the beginning of 1853, there was still no sign of any concerted international effort to get his proposal off the ground and Maury was becoming impatient. Bypassing government channels, he decided to make contact directly with leading meteorologists, hoping they, at least, would jump at the chance to participate in this unique international initiative, and persuade their governments of its merits. In January of that year, he wrote to the observatory on Mauritius, offering the tidy sum of £200 for a copy of its maritime meteorological observations. The Mauritius Observatory was an internationally respected institution under the directorship of Professor Charles Meldrum, another eminent meteorologist engaged in the investigation of tropical storms. Meldrum, however, with typical British restraint, could not be bought by Maury's American largesse, and elected instead to continue to send all his data to the Admiralty.

Luckily for Maury, other meteorologists were much more receptive. Professor Quetelet, Director of the Belgian Observatory, responded enthusiastically, even offering to host the conference, an offer Maury gratefully accepted. Russia, too, was eager to participate. The St Petersburg Observatory of Meteorology and Physics, with its Director Theodore Kupffer, was ahead of its time – an official meteorological observatory operating quite independently of any astronomical link. Only Britain could boast anything similar, and that was the Kew Observatory, which had been acquired by the British Association as a centre for the standardisation and calibration of meteorological and magnetic instruments. Elsewhere, meteorologists were still attached to astronomical observatories – Christoph Buys Ballot in Holland, Heinrich Dové in Prussia, for example – while in France, newly emerged into the Second Empire under its emperor Napoleon III, the Director of the Paris Observatory, D.F.J. Arago, ruled supreme over all meteorological matters. Although not equal in enthusiasm, they did all offer support, giving Maury the confidence to set a date for his conference, in Brussels, for August 1853. The invitations went out. At last, he had what he wanted – an international meteorological conference, albeit purely maritime, and with him, not Britain, taking the lead.

Indeed, far from being in the forefront, Britain was proving laggardly in taking any action at all. Despite the Royal Society recommendation, the prospect of any cooperation with the Americans had fallen foul of the funding problem and the whole project had lapsed into abeyance. Maury's carefully worded invitation to attend an International Maritime Conference to discuss a uniform system for the collection of meteorological data at sea was sent in April on behalf of the Secretary of the US Navy to the British Envoy in Washington, a Mr J.F. Crampton. Crampton duly dispatched the invitation to his superiors in the Foreign Office, who speedily washed their hands of it and sent it on to the Admiralty and also to the Marine Department of the Board of Trade, where it arrived in May, set to languish unheeded.

The Board of Trade's Marine Department had been set up just three years earlier, in 1850, with former arctic explorer Capt Frederick W. Beechey at its helm. Its formation came as a result of an increasing concern over safety at sea, a concern fuelled to some extent by the research into storms by Reid and Redfield, and which was building to the extent where the government felt itself under pressure to address the problem. The Admiralty had the safety and administration of naval ships under its remit, but the numerous merchant fleets came under the control of no official body, so in 1850 a new Mercantile Marine Act was passed, charging the Board of Trade with the superintendence of matters relating to the British mercantile marine and the new Marine Department was created. It was, therefore, a sensible destination for Maury's invitation, even if it generated no reply.

It was fortunate for Maury that reaction from elsewhere had been faster and more positive. The list of countries willing to send a representative was huge and growing – France, Denmark, Russia, Norway, Sweden, Portugal, Spain and the Netherlands. Maury himself would cross the Atlantic to be present as the US delegate. By the end of July, arrangements were nearing completion – and still there was no word from Britain.

By contrast, in France there was no question of whether to attend; discussion centred on who should go. Arago headed a commission of the Academy of Sciences (the French equivalent of the Royal Society) specially convened to select the right person, and, since a seaman rather than a scientist was appropriate for this conference, they chose a hydrographer from the Ministry of Marine. And even in Britain, despite the apathy of government officials, Maury's initiative was receiving a great deal of support from some very influential people. In April, even before the official invitation had actually been received, Lord Wrottesley, a keen meteorologist, future President of the Royal Society and friend of Major-General Burgoyne, was on his feet in the House of Lords delivering a long and passionate appeal on behalf of Maury and his conference proposals. He might have been speaking into thin air for all the notice taken of him by the government, and more specifically by the Admiralty and the Board of Trade.

The date of the conference was approaching fast when, in July, Sir Robert Inglis, MP, finally had the chance to speak in the Commons, a detailed and reasoned argument in favour of Britain's participation. But with the lukewarm attitude prevailing in Whitehall, his efforts produced no results. Scientists and meteorologists were becoming desperate. If Britain did not send a delegation, then the conference would go ahead and make decisions concerning the future of meteorology, leaving Britain with no say in developments. It could not be allowed to happen. Three days after Inglis's speech, a high-powered deputation, including Inglis, Wrottesley and the eminent Edward Sabine himself, descended upon Sir James Graham, First Lord of the Admiralty. Reluctantly, Graham bowed to their demands and appointed a delegation of two representatives – a very unenthusiastic Beechey from the Marine Department of the Board of Trade and, in a surprise move that smacks of the influence of Sir John Burgoyne, a totally delighted Captain Henry James, superintendent of the Royal Engineers' meteorological observers. They went off to Brussels with Graham's strict instructions echoing in their ears: to ensure that Britain's interests were met, but to sign up to nothing that would cost any money.

By now, time was so short that the two had to drop everything they were doing, and rush to pack. Maury's trip was a much more leisurely sail across the Atlantic to England, accompanied by his wife and four very excited young ladies (his daughters Betty and Diana, his niece Ellen Herndon and Ellen Maury, the daughter of his cousin John, Mayor of Washington, DC), off to get their first taste of the sophistications of European society. They were met at Liverpool by Lord Wrottesley and escorted to Wrottesley Hall – the girls enthralled at the prospect of staying in the very house, once called Whiteladies, where a fugitive King Charles II had hidden from Cromwell's forces after the battle of Worcester. During their week in Britain, they were to meet Charles's descendant Captain FitzRoy and, no doubt, Colonel Sabine, Sir Robert Inglis and others who had so vigorously campaigned in support of Maury's initiative. Then they crossed to Paris to be entertained by the prominent astronomer Le Verrier, followed by a grand triumphal progress across Europe, terminating finally as the guests of Quetelet in Brussels. For Maury, what a contrast to Washington this all must have been. Here there were no bitter wranglings, no reactive naval superiors to fight against. Suddenly, he was a celebrity, fêted for his wind charts, and for bringing about the coming international conference that he was poised to lead.

The conference was held in the official residence of the Belgian Minister of the Interior, M. Piercot. Amid the panelling and portraits, the delegates gathered, best navy worsted brushing against gold-trimmed epaulettes as the men mingled and conversed in a mix of languages, French prevalent, though with English predominant, notably in the soft transatlantic tones of the American contingent. Apart from the host, M. Quetelet, the scientists were absent, replaced, since this had evolved into a purely maritime exercise, by naval representatives – Lahure, Director of the Belgian Marine, Rothe from the Danish Depot of Charts, Delamarche of the French Marine Hydrography Department, Jansen and Ihlen of the Netherland and Norwegian navies, while Lieutenant-Captains de Mattos Corres and Gorkovenko came from Portugal and Russia, with First Lieutenant Pettersson representing Sweden. Maury must have sought the English voices of Captain Beechey and Captain James with anticipation – James perhaps with some trepidation, since it was his pamphlet of instructions to his own observers that had formed the basis of this entire conference, and he might well prove antagonistic to the naval appropriation of the British Army's first initiative; and then Beechey, with whom Maury must have itched to discuss the curious 'bottle charts' that he had unearthed in the archives in his early days at the Depot of Charts. Beechey was a renowned naval surveyor, whose authoritative navigational charts would have been standard issue to foreign shipping, including the USA. His 'bottle charts' were more of an oddity. In an effort to trace the effects of the great currents of the Atlantic, Beechey had once tossed from his ship a series of 100 bottles, at various locations in the Atlantic. His chart showed where these bottles had been retrieved – some in Europe and some in Africa, while most had found the Gulf Stream and had whirled across the North Atlantic to be washed up on the coasts of the British Isles.

Maury would have searched for the two in vain, though, since they failed to arrive in time for the conference's first day. They missed his big moment when he rose to give the opening address to the delegates.

The object of our meeting . . . is to agree upon a uniform mode of making nautical and meteorological observations on board vessels of war. I am already indebted to the kindness of one of the members present, Lieutenant Jansen of the Dutch navy, for the extract of a log kept on board a Dutch ship of war, and which may be quoted as an example of what may be expected from skilful and carefully conducted observations. In order to regulate the distribution of the charts, which the American Government offers gratuitously to captains, it would, in my opinion, be desirable, that in each country a person should be appointed by the Government, to collect and classify the abstracts of the logs, of which I have spoken, through whom also the charts should be supplied to the parties desirous of obtaining them . . .

Allow me to add that we are taking part in a proceeding to which we should vainly seek for a parallel in history. Heretofore, when naval officers of different nations met in such numbers, it was to deliberate at the cannons' mouths upon the most efficacious means of destroying the human species. To-day, on the contrary, we see assembled the delegates of almost every maritime nation, for the noble purpose of serving humanity by seeking to render navigations more and more secure. I think, Gentlemen, we may congratulate ourselves with pride upon the opening of this new era.

Grand words for a big occasion – peacetime international cooperation on a scale never attempted before. It remained to be seen whether such high ideals could be attained. Surprisingly, with such a mix of personalities and nationalities, the conference was unusually amicable and constructive. Discussion covered a range of detailed points concerning the means, number and frequency of the meteorological readings to be taken on board both naval and merchant ships. Henry James (arriving eventually with Beechey on the evening of the first day) slotted in without a qualm, the only army man in the naval gathering, and joined a subcommittee discussing the format of the proposed logs, generously conceding amendments to his own original design. By the end of the conference, on 8 September, uniform methods of measurements had been agreed, along with the prescribed form of log for recording them. Maury was jubilant at the success, and while the other delegates went home to plan the means by which the observations could be gathered, he envisioned the increasing accuracy of his wind charts and how their numbers would expand to cover all the world's oceans.

As senior representative from Britain, the other English-speaking nation along with the USA, Beechey, the most reluctant delegate, remained in Brussels with Maury to write the report of the conference. As for Henry James, he left Brussels a happy man. On his return to his home in Edinburgh, he immediately got out his pen and paper and wrote of his experiences in a letter to Burgoyne. 'The Science of Meteorology can now be successfully cultivated,' he euphorically declared.

Beechey, though, was probably not quite so enchanted. Maury's wind charts might be coming for free, but only in exchange for much effort and expense in the issue of instruments and the organisation of returns. None of the delegates had, in fact, any power to commit his home government to any specific action, yet if all other countries were seen to be participating, how could Britain opt out? Indeed, he himself had written in his own report:

The Conference, having brought to a close its labours with respect to the facts to be collected and the means to be employed for that purpose, has now only to express a hope that whatever observations may be made, will be turned to useful account when received and not be suffered to lie dormant for the want of a department to discuss them . . .

Charged, before he went, with agreeing to nothing that would result in any expenditure, Beechey had put his name to a document that essentially committed Britain to the establishment of an official Government Meteorological Office.

SIX

Who will be the Weatherman?

In October 1853 Turkey declared war on Russia. It may seem that both Turkey and Russia were somewhat remote from the autumnal London to which Beechey had returned, but the likelihood was that neither would remain distant for long. When news of trouble flaring on the Russian–Turkish borders first filtered through in the summer, Britain and France, with interests in the Middle East to protect, had mustered a combined fleet and sent it into the Dardanelles. Now, with the Turkish declaration against Russia, Britain needed to prepare for inevitable full-scale military action. An insignificant gathering of naval captains talking to each other of winds and the weather, which already occupied a ranking hovering just above zero in the priorities of the British government, plummeted even lower. When he returned from the conference, Beechey filed his report and turned his attention to dealing with the threat of a naval war in the eastern Mediterranean and the subsequent disruption to shipping.

For Henry James, however, war or no war, nothing could be more important than the threshold on which stood the science of meteorology. He determined to take steps to progress it. Beechey had hardly tucked his feet back under his desk in Whitehall when James was badgering him about the arrangements he was making to implement the Brussels agreement. James himself was already dreaming up a scheme to incorporate land observations into maritime returns, and confidently told Beechey that he expected another conference on the subject to take place very soon. James, it should be remembered, had his own interests in meteorology, which revolved around the work he had done for Burgoyne in supervising the collection of readings from the various Royal Engineers' establishments in the Colonies. As far as he was concerned, Maury's successful approach to Europe, and the agreement on maritime observations, were only a start. Maury had already been in contact with him, keen to secure the soldier's support for a land conference, and it seemed to him that it was time for Britain to stop dallying and to take a lead in the field. Realising he needed a 'proper' scientist on his side, and with the Royal Society having already shown its colours for the maritime observation lobby, James scouted around for another likely advocate and alighted upon Professor George Airy, the Astronomer Royal. Airy seemed a sound candidate, since he had already proved his worth as a keen 'weather man'.

Five years earlier, in 1848, the late summer had been extremely wet and farmers were issuing doom prophesies for the harvest. The country was beginning to descend into panic. In an effort to help, the *Daily News* published each morning a statement of the state of the wind and the weather. The newspaper had enlisted the help of twenty-nine stationmasters in various parts of the country, who at nine o'clock every morning gazed skywards, watched windvanes, gauged the rainfall and sent their reports, via the London train, back to the editorial offices of the *Daily News*. Whether it helped the harvest is not proved, but it certainly excited the readers, because when the reports ceased at the end of October, they begged for them to continue.

Professor Airy was one such reader. However, his interest was not just casual. He wanted to use the reports to plot readings onto weather maps for further investigation. The newspaper was more than keen to reinstate the weather statements, especially when Airy offered to send out his long-suffering but highly efficient and enthusiastic meteorological assistant, James Glaisher, on a gruelling 'station-crawl' to inspect the stationmasters' crude instruments and improve their skills in using them.

James Glaisher could lay claim to being Britain's first salaried, specialist meteorologist. In his role as superintendant of the Magnetic and Meteorological Department at the Greenwich Observatory, a post he had occupied since its foundation in 1840, he organised meteorological observations and investigations, and took charge of the accuracy of the instruments involved. Glaisher was forthright, dedicated and a stickler for detail. His love of precision instruments came early in life – perhaps even through his genes – since his father, also

James, was a watchmaker who ran his own workshop in Rotherhithe. This east London docklands area was not a vast distance from the open spaces of Greenwich, and one of young James's early friendships was with William Richardson, who worked at the observatory on the hill in the park. His first sight of the observatory, with its telescopes, its clocks and its vast range of scientific instruments, must have seemed like wonderland to the lad fresh from a watchmaker's workshop. It certainly made an impression, since James Glaisher was never happier than when he had an instrument in his hand and a purpose for using it.

At the age of 20, James gained a position with the Ordnance Survey and, with his sextant and theodolite, he climbed first to the summit of Ben Corr in Connemara, and then The Keeper in Limerick, to assist in the triangulation of Ireland. A modern theodolite is a small, light and easily transportable instrument, but the early nineteenth-century equivalent was an unwieldy contraption, and, once it had been dragged up a mountain, there it would have to stay until the job was done – and its operator with it. Up in the clouds, blasted by the winds and battered by the driving rain, Glaisher spent long, lonely watches taking trigonometric measurements, and observing the weather. 'I was thus led to study the colours of the sky, the delicate tints of the clouds, the motion of opaque masses, the forms of the crystals of snow.'

While his exposure to the mountain airs instilled in him a fascination for meteorology, it also inflicted a severe attack on his constitution, which had him descending the mountains for more congenial employment in the fenland county of Cambridgeshire. Here he swapped his theodolite for a telescope and pointed his sextant at the stars as assistant to Professor Airy, who was then with the Cambridge Observatory. Astronomical observations would frequently be curtailed by curtains of cloud that suddenly arrived to obscure the stars, and Glaisher was keen to know 'the cause of their rapid formation and the processes in action around them'. In 1835 Airy was appointed Astronomer Royal and transferred to Greenwich, taking Glaisher with him and thus bringing him back home to the observatory that had first inspired him.

Glaisher's duties were meteorological as well as astronomical, and it was Airy who recognised the importance of his meteorological role when he created the Greenwich

Observatory's Meteorological Department and appointed Glaisher as its head. Glaisher's enthusiasm soon surfaced, as it had in Ireland, into personal investigations, and one of his first was into the effect on the 'dew point' of stored heat being radiated back from the ground at night. The dew point is the temperature at which air, as it cools, releases into droplets the moisture it took in during the heat of the day. True accuracy of its measurement required a series of temperature readings taken at ground level each evening just as the dew began to form on the grass. Not one to shirk from bodily discomfort in the pursuit of his science, Glaisher spent chilly night after chilly night, prostrate in Greenwich Park, waiting with his thermometer for that telltale moment when the first dewdrop appeared. He soon discovered that, even at ground level, research could be hazardous to health, and the experiment had to stop when severe rheumatism set in. He also discovered that none of the thermometers available at the time was capable of registering the nuances in temperature he needed, so the watchmaker's son set about designing his own.

Airy's professed interest in weather maps was more properly that of James Glaisher. Even before the professor lent Glaisher's services to the *Daily News*' network of stationmasters, Glaisher had established his own group of eager amateurs, dedicated to taking meteorological measurements and posting them off to Greenwich, where they could be plotted onto maps. By 1850 interest in meteorology had grown so rapidly that there was a demand for an association at which amateur meteorologists could meet and exchange observations and ideas. On 3 April, in the library of Hartwell House, near Aylesbury, Buckinghamshire, the home of a Dr John Lee, a gathering of ten such men took place. Together, they inaugurated the British Meteorological Society. Samuel Charles Whitbread, of the famous brewing family, was elected its first President, while James Glaisher, also present that evening, became Secretary, a post he was to hold for the next twenty-two years. News of the society soon crossed the Atlantic, and winging back came a request from none other than Joseph Henry of the Smithsonian Institution, asking Glaisher's help in designing a standard reporting format for his network of North American observers. Glaisher posted off a copy of the form his stationmasters used, and happened to mention to Henry that it would be a good idea if the captains of ships crossing the

Atlantic could be persuaded to complete similar forms, providing observations of the weather at sea. Poor Glaisher – this was precisely at the time when Maury's first North Atlantic pilot charts had been published and he was apparently unaware that he had stepped unwittingly into the storm of bitterness raging between the two American rivals. Not surprisingly, Henry neglected to take up Glaisher's suggestion, and it appears to be the first and last time that Glaisher took an interest in maritime meteorology.

With such an active land-based network operating under the auspices of the Royal Observatory, it was perfectly logical for Henry James to reckon he could persuade Glaisher and Airy to support his thrust for a land conference. It was not the first time Glaisher had cooperated with James. Together they had discussed the instructions that James had distributed to his teams of Royal Engineers' weather observers, and no doubt it was Glaisher's ideas, based on his stationmaster chain, that had gone into the design of the weather stations to be constructed and the instruments to be used. James was right in his assumption. In reply to his approach, Airy wrote: 'In the whole world there is no science so overwhelmed with undigested facts as meteorology.'

While the battle for the land conference was being joined, another thorny matter was waiting to be thrashed out. That there was a need for a new 'Meteorological Department' was not in question – all these Brussels-instigated maritime observations had to be coordinated and collated somewhere – but exactly where and how and by whom was very much open to debate. If there was to be a new department, then Beechey wanted it, not so much for the honour of running the department itself, as for the advantage of obtaining control of its budget, which would swell the coffers, and thereby the importance, of his Marine Department. Henry James, too, seems to have assumed that, since Beechey was the government representative at the conference, he would be the prime mover in organising the implementation of its recommendations. As for James's new ally Airy, he thought it unimportant which government body had nominal control of the new department; what really mattered was the man who would actually run it. He declared to James that 'I have no objection to undertaking it'.

Beechey, however, seemed to be diverting insufficient energy into the formation of a Meteorological Department, and Henry

James was becoming frustrated. Going straight to the top, he told Edward Cardwell, MP, President of the Board of Trade, that he wished to see the new Meteorological Department as part of the Board of Trade, and that he had the ideal candidate for the supervisory role in the person of Professor Airy. Would Mr Cardwell agree to speak to him? When the President did agree to the interview, a jubilant James shot off a message to Beechey, telling him of this new development.

In the meantime, however, a team of heavyweights was manœuvring in opposition, and the biggest among them was Colonel Edward Sabine. Sabine had watched with interest the results coming out of the conference and had wasted no time in setting in motion his own preparations for the work to be done. It was, of course, partly on his say-so that the Royal Society had recommended the rejection of land-based observations, regarding the new department as a purely naval concern, rightly administered by the Admiralty. To Sabine, the ideal man to run it would not be a scientist. The job required someone who could organise, who could speak to ships' commanders on their own terms, who was conscious of the problems involved in taking observations at sea, who appreciated the aims of the exercise, but yet whose insufficient knowledge of the science behind it would prevent him interfering by imposing his own ideas. In short, a conscientious, out-of-work, ex-naval officer with an interest in meteorology would fit the bill beautifully. And Sabine had in mind the ideal man – Captain Robert FitzRoy.

SEVEN

Surveying the Seas

Robert FitzRoy's career had led him through a series of peaks and troughs as dizzying as any storm-slashed sea he had ever encountered. The fact that he was now, in 1853, languishing penniless – or what passed for penniless among the Victorian upper classes – did not mean that he had not seen wealth, fame and glittering successes in his time. However, it is true to say that his very first solo command was fated to end in ignominy. The lad was only 5 when he sneaked out of the house with an old washtub, dragged it down to the lake at the bottom of his garden, launched himself into it and promptly overturned.

Even at that early age, the attraction of water and sailing was irresistible. He was born in 1805, the year of the Battle of Trafalgar and the death of Admiral Lord Nelson. Nelson's funeral was an occasion of great public mourning, amid lavish pomp and ceremony, while the population celebrated the glory of Britain's naval supremacy. Like young Matthew Maury reading of the adventures of his sailor brother, Robert would no doubt have thrilled to tales of sea battles and the epic adventures of his naval ancestors and other maritime explorers such as the legendary Captain Cook – all fuel to the imagination of an eager, enquiring youngster.

With just one year between them, Maury and FitzRoy were of the same generation, yet a world apart in both location and situation. In contrast to the Maurys' Tennessee shack, the FitzRoys' home was no mean hovel. Wakefield Lodge, in the village of Potterspury, bordered the former Roman road of Watling Street, where it followed its ancient line of progress from Buckinghamshire into Northamptonshire. Built as the hunting lodge of the Dukes of Grafton, it stood on rising ground above two landscaped lakes, showing its stately Palladian face to the 30,000-acre park. Within its walls it boasted a ballroom, dining room and withdrawing rooms fit to cater for the evening entertainment of the Duke, his guests and their ladies. But the old duke had died in 1804, and the new duke, George Henry FitzRoy, fourth Duke of Grafton, had no need for a vast second house that lay empty for months on end. So his younger brother, Lord Charles FitzRoy, with wife Frances, daughter, also Frances, and three young sons, Charles, George and Robert, left the family mansion of Ampton Hall in Suffolk and came in 1809 to take up residence.

Lord Charles FitzRoy could never lay claim to being a man of note. After an unremarkable though competent career in the Army, he now had, as member for Bury St Edmunds, a parliamentary seat upon which he sat only infrequently, and was content to live quietly within his adequate means, riding to hounds, tending his garden and overseeing his estate. Not so his sons – Charles, the eldest, with a reputation as a fast-living, blustering aristocrat, was to rise to the position of Governor of New South Wales, whereas Robert, his young half-brother, was destined to shine as one of the Royal Navy's most adventurous and capable commanders. Perhaps Charles FitzRoy's greatest gift to his youngest son was to teach him the wonders of the weather glass, showing him how to follow the rise and fall of the mercury in the barometer in the hall in conjunction with the skies beyond the windows.

Robert came from a line of illustrious antecedents. 'FitzRoy' means literally 'son of the king', and the dynasty was founded when the Merry Monarch, Charles II, and his mistress Barbara Villiers, produced an offspring, a royal bastard given the title Duke of Grafton. Charles II's father, Charles I, had been a highly principled aristocrat whose determination to cling doggedly to his beliefs lost him his head. His descendant, Robert FitzRoy, appears to have inherited this same selfless sense of duty, a trait that would not always stand him in good stead. The impulsive, buccaneering spirit of the second King Charles, though, is evident in a line of FitzRoys who, including the first Lord Grafton, were fearless naval captains, achieving command young, and dying young in action. Following such flamboyant forebears, it was hardly surprising that Robert was destined for a career in the Navy.

Usually, boys of his social background would have been sent away to sea at 12 or 13, pitched straight from home comforts into the realities of a lifestyle immortalised by Captain Marryat in *Mr Midshipman Easy*. For Robert, though, this practice was dispensed with in favour of a prior specialist education at the Royal Naval College in Portsmouth. In February 1818, not quite 13 years old and already a veteran of Rottingdean and Harrow schools, Robert arrived at the doors of the college, clutching his bag and as much courage as he could muster.

Whether young Robert thought so at the time or not, the college route was a happy choice for him, since formal learning suited his temperament. It would be said of him in the future that 'I never in my life met a man who would endure nearly so great a share of fatigue. He works incessantly, and when apparently not employed, he is thinking', and this was equally true of FitzRoy, the student, applying himself almost to a state of exhaustion, determined to excel. Along with the usual classical subjects, the college taught 'drawing' (i.e. draughting), naval architecture and gunnery and introduced him to the sciences of navigation, astronomy and mechanics. FitzRoy's mathematics, weak on entering, had improved to such an extent that by the time he left, only eighteen months later, he had won the Mathematics prize. There is in the archives of the National Maritime Museum in Greenwich a certificate of mathematics awarded to a John Jervis Tucker in October 1817, just two years before FitzRoy. It says that he

> has been examined touching his knowledge of the elements of mathematics and the theory of navigation; more particularly in the necessary parts of arithmetic, in the mode of observing and calculating azimuths, amplitudes and the variation of the compass, in the calculation of tides, the various modes of ascertaining latitude, as well as the simple and double altitudes of the sun, as by the altitudes of the moon and stars; and the finding of longitude by chronometer and lunar observations.

So, unlike Matthew Maury struggling to teach himself on the job, Robert FitzRoy started his career with the best education that money and privilege could provide. This was the post-Nelson era when the age of sail was at its peak, yet life on board a sailing ship was basic, dangerous and hard — especially hard when you find yourself the fancy college boy, butt of the teasing and downright antagonism of the old hands. But it is human nature to love to be loved. So when, just 19 and newly promoted to lieutenant, Robert arrived on board the *Thetis* in 1824 and found young Bartholomew Sulivan, four years his junior and a college boy himself, looking up to him with admiring eyes, he was only too happy to take him under his wing.

The *Thetis*'s duty might well have seemed tame, patrolling the rocky coast of Cornwall on the lookout for smugglers, but it gave the pair ample opportunity to experience sailing in adverse conditions and to study landfalls and currents, winds and the course of storms. After three years of patrol, the ship was posted to the South American station. One cloudless night in Rio de Janeiro, the *Thetis* was struck by lightning. No warning, no rain, just a bolt from nowhere, which struck the top of a mast, and instead of shooting down by the shortest route, twisted round and around and came out by each iron loop, even causing one to explode. Then came the strange light known as St Elmo's fire, which appeared at the masthead and the tip of each yard arm. St Elmo's fire is a 'brush' electricity, formed in the highly charged atmosphere of a storm. But FitzRoy did not know that. Impelled to discover for himself what this startling phenomenon might be, he climbed out onto the yard arm and thrust his hand into the light. It did not burn or tingle – in fact he felt nothing. Then suddenly the light disappeared as quickly as it had come, and the ship was deluged by rain. The *Thetis*, as well as FitzRoy, survived the incident undamaged.

Despite finding little real action in which to show his mettle, FitzRoy nevertheless succeeded in attracting the attention of the Admiral of the South American station, Admiral Sir Robert Otway, and was transferred from the *Thetis* to the *Ganges*, as Otway's flag lieutenant. Fate would have it, therefore, that he was on the spot in Rio when the surveying barque, *Beagle*, limped into port carrying a sorry tale. The *Beagle*'s captain had committed suicide, by all accounts driven to madness by a long stint of surveying amid the barren, hostile coasts at the tip of South America. The *Beagle* was on duty as part of a two-ship Royal Naval survey expedition to South America, led by Captain Philip Parker King (son of Philip Gidley King, Governor of New South Wales) from on board the *Adventure*. King had returned to Rio in order to seek the permission of Admiral Otway to appoint Lieutenant Skyring, temporary captain of the *Beagle*, to full command. It is difficult to know who was more surprised, King or FitzRoy, when the Admiral appointed FitzRoy instead. Suddenly, the young lieutenant was thrust not only into command, but into the role of naval surveyor.

Surveying the seas had long been a peacetime occupation of the Royal Navy. Captain Cook, of course, had made brief surveys of the new lands he discovered, and George Vancouver, in 1790, while attempting to find a navigable river to connect Canada's west coast with the Great Lakes, had produced valuable charts of the coast of British Columbia. The outbreak of the Napoleonic Wars had put a stop to scientific and survey work, but afterwards, when it became apparent that more ships had been sunk by running aground than had been destroyed by enemy guns, surveying took on the role as one of the prime duties of the Navy. Spare ships – and there were many after the war – were recommissioned as survey vessels and went out to chart the world. In view of its increasing importance, the Hydrographer's Office was reorganised and restructured and Captain Francis Beaufort, its newly appointed chief, became established as the most efficient and respected keeper of charts in the world.

Beaufort was himself a skilled surveyor, who had spent eighteen months charting the eastern Mediterranean until June 1812, when he was shot by a Turk and wounded out of active service. Beaufort's skill as a draughtsman was unsurpassed. His charts, along with sketches of the shoreline, views and sailing directions set the standard for others to follow, and the magnificent illustrations of ancient sites he produced on his trips ashore were hailed as masterful, as much of a wonder as the ruins themselves. Beaufort had long recognised that, besides charts, sailors needed information on such items as currents, climates, safe anchorages, where to find fresh water, what native animals and vegetables were available to replenish provisions – and, naturally, the wind. Recording the wind on the ship's log had always presented a problem, with descriptions of wind strengths being as varied and inventive as the individuals who noted them. To solve the problem, Beaufort conceived a standardised system for allocating a simple number to a given wind speed.

To be fair, the idea was not new. Robert Hooke had used a numerical wind-speed scale in the seventeenth century, while in the eighteenth various scales appeared on the logs of serious weather observers such as John Dalton and his friend Peter

Crossthwaite in Cumbria. The noted English civil engineer John Smeaton created a scale of wind speeds for use by millers as a guide to the amount of canvas to carry on their sails and Alexander Dalrymple, hydrographer to the East India Company and friend of Smeaton, instantly recognised the worth of a similar scale for use at sea. In those days, it was the merchants, not the Navy, who hankered after accurate charts – a sunk ship being a cargo lost and an investor's fortune with it. Dalrymple had a fleet of dedicated survey vessels out charting the Company's routes and in 1779 he issued them all with a pamphlet detailing his wind speed scale. It was on one of Dalrymple's survey ships that, ten years later, a young Francis Beaufort first went to sea.

Confrontation with the French drew both Dalrymple and Beaufort from the East India Company to the Navy, Dalrymple becoming the first Naval Hydrographer and Beaufort a lowly midshipman. As a civilian, Dalrymple faced the devil of a job – naval officers shunned him, ignoring his pleas for ships' logs to be forwarded to his office. Apart, that is, from Beaufort. In 1805, in command of his first vessel, Beaufort agreed to cooperate, and, in exchange, a grateful Dalrymple presented him with a copy of his East India Company scale of wind speeds. Beaufort adapted Dalrymple's scale, adding an extra point to it and, inspired perhaps by Smeaton's windmills, a description of sea states based on the sails a ship could carry in a particular wind. Originally published by him in 1806 (though not officially adopted by the Navy until 1838), his scale numbered from 0, dead calm, to 13, hurricane.

Despite its predecessors, Beaufort's was to be the scale to win universal acceptance, and this was entirely due to Beaufort's efforts, when he became Naval Hydrographer himself, in insisting on its use. His scale as it is known and used today is basically the same, although states of sails have been replaced by states of sea or certain signs on land – force 4, moderate, for example, is recognised by numerous white caps on the sea or by loose paper, dust and leaves being blown and small branches moving on the trees. Like Luke Howard's cloud classifications, the beauty of the Beaufort scale is that it needs no instruments yet it enables a vital element of weather to be accurately gauged and a comparative record noted by different people in disparate places. Where would Maury's wind charts have been without it? How could the Brussels Conference ever have defined an international standard for wind speed measurements? Beaufort enabled meteorology to take a major step forward.

The maritime peace following the Napoleonic Wars allowed for the expansion of world trade on a massive scale. International trade needed mercantile fleets to serve it and merchantmen from all nations streamed across the oceans and swarmed up the coasts of foreign lands. Not just lives but money was at stake now and a ship's master would attempt to acquire whatever charts he could for his voyage – be they Spanish, Dutch, Portuguese, French, or British – and the Royal Navy sought to provide the best. Surveying, though, was not the pleasant peaceful occupation it might sound. A good surveyor was not simply a good draughtsman. He had to be a superb navigator, too. With only his skill to guide him, he had to penetrate every river mouth and bay, round every headland, measure low shoals and high cliffs and record the contours of every jutting rock, above and below the surface. The surveyor's enemies were more deadly than any he might meet in battle. In 1821–5, when Captain William Fitzwilliam Owen made his great survey of Africa, the 300 charts he created were said to be coloured red with the blood of men. For his crew, malaria was the foe. The cause of the curse was unknown – it was thought to be in the vapours of swamps or on the mudflats of rivers and lakes – and Owen's crew succumbed in numbers that would have been even greater had they not come across a Frenchman in Madagascar who prescribed quinine as a cure.

In South America, it was the weather and the terrain that waited to claim lives. The Spanish had recently relinquished their colonies, opening up, along with vast new trading possibilities, hundreds of miles of treacherous, badly charted coasts, including the notorious Cape Horn and the southern land of Tierra del Fuego with its tortuous cliffs, narrow passages, savage winds and sudden storms. It was in this hostile region that Captain King on the *Adventure* had spent two harrowing years, and it was here that Commander Pringle Stokes of the *Beagle*, driven mad by the stark desolation and the intense loneliness, had placed his pistol at his temple and pulled the trigger.

At only 23, FitzRoy, as Stokes's replacement, was very young to be given his first command. And it was a daunting prospect. Skyring, quite naturally, would be resentful, the crew despondent, and, what is more, the ghost of the former captain was said to haunt the ship. FitzRoy was haunted, too, by his own private ghost. Stokes's grizzly end would have conjured the horror of a family tragedy played out in public, not many years before. His mother, Lady Frances Stewart, was the sister of Lord Castlereagh, the Foreign Secretary who had represented Britain at the Congress of Vienna following the defeat of Napoleon. Castlereagh was self-contained, highly strung and, like his nephew, an indefatigable worker who did not spare himself. Famously, he fought a duel with George Canning, the future prime minister, survived it, then committed suicide in 1822, believing he was being blackmailed for homosexuality. Robert FitzRoy was 17 when Castlereagh died and he must have been torn apart by the public outcry and the fact that his uncle had committed the ultimate Christian sin.

With a hostile crew to mollify and fearful that a possible inherited suicidal tendency might drive him to the same fate as his predecessor, it must have taken all the young commander's stoic reserve to stand confidently on deck and calmly inch his vessel out to sea. Even the weather conspired to threaten his command. Scarcely had they cleared the harbour when the barometer dropped suddenly to a very low 28.5 inches, but Robert FitzRoy, budding meteorologist and keen observer, neglected to take due note. In the storm that followed, the *Beagle* lost both top masts along with many of her spars and she nearly capsized. It was a hard test, and in the circumstances he could be forgiven for failing it. But FitzRoy never could forgive failure, especially in himself.

Robert FitzRoy took to surveying with flair. Mapping required painstaking and accurate measurement, intricate calculation, plus great skill in drawing, in all of which he excelled. The waters around the jagged tip of South America are some of the most dangerous on earth, where sudden squalls, or williwaws, can blow up out of a clear blue sky, with a force that would scuttle a ship, let alone the small boats needed for close surveying. The acute senses of a skilled weather man were needed to keep a ship safe in such waters, and FitzRoy's foresight saved the ship from disaster on many an occasion. FitzRoy did not baulk at danger, and by venturing out personally in the boats and sharing the same risks as his crew he soon re-energised and inspired them. His own motivation came not only from his sense of duty but from the knowledge that accurate charting would be the salvation of future generations of seamen.

By August 1830 he was returning to London, the tour of duty successfully completed, although with much of the country still uncharted. The Admiralty, though, was not prepared to fund a further mission to South America to complete the survey. Dismayed, FitzRoy found a champion in Captain Beaufort, now Naval Hydrographer, who campaigned for this important work to be continued. The *Beagle* was refitted to FitzRoy's specification, controversially rigged with lightning conductors and equipped with all the latest instruments, including a set of twenty-two chronometers to achieve the second part of the expedition's purpose, which was to establish a reliable set of longitudinal readings across the entire globe. The crew were selected, Sulivan among them, and added to their number, in order to allay the extreme loneliness of his position, a companion for FitzRoy. This person had to be a scientist, someone who could benefit from the voyage, with means sufficient to pay his own way, an equal in social standing and, most importantly, a man of easy disposition. Charles Darwin was eventually chosen – and how wise yet fatally ironic this choice would prove as future events unfolded.

FitzRoy left England in December 1831 with a full set of Admiralty orders for the voyage. They covered every detail of what he was expected to achieve in his five-year voyage, how, when and where he would achieve it, plus a long list of desirable extras, among which were instructions in compiling meteorological registers, which 'may be of use – but only if steadily and accurately kept'. Once at sea, however, he was on his own. While he had great respect for the objectives of his mission, he held in scant regard those details that were at odds with the situations in which he would find himself. Only a few months into the survey it was obvious to FitzRoy that the task expected of him was totally unrealistic. Simple logistics were against him for a start – it took a whole month just to sail to the nearest naval supply port for reprovisioning – and then there was the sheer extent of deeply indented coastline, the rocks and shoals that all needed to be charted in order to make the coast safe for shipping. Either he reduced the accuracy of his charting, potentially missing safe anchorages and dangerous obstacles, or he abandoned part of his task, or he obtained an additional vessel. To FitzRoy, the mission was all, so, against instructions, he bought and had fitted a supplementary schooner at his own expense. He figured that, since the Lords of Admiralty had never before sent just one ship on such a lengthy surveying mission in such a remote part of the world, they would agree to foot the bill. They did not. 'Their Lordships do not approve of hiring vessels for the service and therefore desire that they may be discharged as soon as possible.' Their refusal dealt him a blow, healthwise as well as financial. He wrote to Beaufort:

> My schooner is *sold* . . . The charts are progressing slowly . . . I am in the dumps – It is heavy work . . . Troubles and difficulties harass and oppress me . . . Having been obliged to sell my schooner . . . – Continual hard work – and heavy expense . . . These and many other things have made me ill and very unhappy.

Darwin took up the story in a letter to his sister:

> the selling [of] the schooner and its consequences were very vexatious, the cold manner the Admiralty (solely I believe because he is a Tory) have treated him, and a thousand other, etc. etc.'s, has made him very thin and unwell. This was accompanied by a morbid depression of spirits, and a loss of all decision and resolution . . . All that Bynoe [the Surgeon] could say, that it was merely the effect of bodily health and exhaustion after such application, would not do; he invalided, and Wickham was appointed to the command. By the instructions Wickham could only finish the survey of the southern part, and would then have been obliged to return direct to England. The grief on board the Beagle about the Captain's decision was universal and deeply felt.

Was FitzRoy about to choose the fatal course taken by Castlereagh, and poor Captain Stokes? Was Darwin to miss the Galapagos Islands, not stumble across their unique wildlife, nor take away from there the specimens from which his theory of evolution would eventually result? Fortunately for FitzRoy, Darwin and the *Origin of Species*, the captain was brought back from the brink by the diplomacy and consideration of his second-in-command, Lieutenant Wickham. 'He has already

regained his cool, inflexible manner, which he had quite lost,'
remarked Darwin with relief, as FitzRoy took command once
more. The *Beagle* left South America for the South Pacific and
the series of staging posts designated for completing her task of
ringing the world with a chain of longitudinal positionings. In
contrast to the excitement they created for Darwin, the
Galapagos Islands were to FitzRoy just another meridian to
measure, another coastline to survey. It was Tahiti and New
Zealand that would feature in his reports, since it was here
that he found himself drawn into internal politics.

Forced upon him had been the delicate task of extracting from
the Queen of Tahiti a large fine imposed for the murder of a
British sailor. A situation that might have proved difficult was in
fact delightful. The Queen and the islanders, warming to
FitzRoy's evident dignity and sense of fair play, rallied round
and paid up cheerfully. In contrast, newly colonialised New
Zealand did not furnish nearly so pleasant an experience.
FitzRoy was confronted by a hotpot of simmering trouble, with
missionaries doling out Christianity to the Maoris on the one
hand, whalers and escaped convicts dishing up rifles and rum to
them on the other, and all presided over by an ineffectual 'British
Resident' with no official powers at all. Frequently appealed to,
yet declining, to sit in judgement in disputes, FitzRoy finally
escaped to the cabin of the *Beagle*, where he poured his thoughts
on the problems of New Zealand into the pages of his journal.
Finally, he sailed for Sydney and thence for home.

The euphoria of homecoming was matched by a wave of
public acclaim. Darwin, who had left as little more than a
student, returned a fully-fledged scientist, his collections,
notebooks and sketches pounced upon by eager naturalists.
FitzRoy was fêted for his superbly drawn charts, which would
be saving sailors' lives for 100 years to come, and was
presented with the coveted Gold Medal of the Royal
Geographical Society. Yet they had come home to a kingdom
in the throes of change. It was 1837, the year a young Queen
Victoria ascended the throne. The country was thrusting
forward towards a booming future and great feats were
expected of her talented subjects: scientists, inventors,
explorers and people such as FitzRoy, the capable and gifted
master of the seas.

Alas for FitzRoy, this was not to be. After the voyage of the
Beagle, he did not serve at sea again. The job had exhausted

him, not so much through its physical exertions as through the mental strains of living up to his own elevated sense of duty while dealing with the frustrating restrictions imposed by the short-sighted, penny-pinching attitude of the Admiralty. Perhaps, also, the attractions of married life were claiming too strong a hold. Mary O'Brien had waited five long years for her sailor beau and they were married just two months after he returned. Mary was a deeply religious woman, and it soon became clear that Robert's own faith was intensifying and veering to the fundamental under her influence. As FitzRoy was cultivating an unshakeable belief in the literal truth of the biblical story of creation, so Darwin was beginning to develop his diametrically opposing theory of evolution of species by natural selection. Five years of friendship rapidly started to crumble as their beliefs diverged and hardened. While Darwin truly regretted the rift, FitzRoy was deeply wounded by it, and the hurt was to fester and emerge to dominate his later life.

Meanwhile, FitzRoy was searching for alternative employment and it was his Londonderry relatives, staunch Tories, who persuaded him into politics. He stood as a Tory for the parliamentary seat of Durham, which he won only after a disgraceful debacle involving a challenge to a duel with a rival candidate. FitzRoy was quick to exploit the power of his new position. He campaigned for improved safety at sea and proposed to the Board of Trade a system of examination boards for ships' masters. Although his plan was not accepted initially, it became voluntary three years later, and compulsory by 1850. As Acting Conservator of the River Mersey, he was given the chance to apply his surveying skills and, devoting to the post a personal involvement way beyond that required, happily headed north to inspect the river, its tides and erosions.

But the Colonial Secretary, Lord Stanley, had other plans for him. The post of Governor of New Zealand had become vacant, and, having been to the place and already involved himself in its affairs, FitzRoy seemed the obvious choice. He sailed off to preside over the most impossible of situations. FitzRoy, the champion of the indigenous native, set to with a will to support the Maoris against the vested interests of white settlers, white traders and white missionaries, all ranged in opposition against him. That he did not succeed was not for want of trying. Without military support, he was backed onto the defensive with his wits his only armoury; and with no financial backing, he was forced to use his own fortune in place of state funds. If it had not been for the ignominy of it, his subsequent recall by order of Parliament might have been a relief. Under extreme pressure, FitzRoy had suffered the sort of stress-related illnesses he had experienced on the *Beagle*, and without his wife's loving support, he might well have succumbed this time. They left New Zealand in November 1845, and it cannot have helped Robert's dented pride to come across, en route back home, his own half-brother, Charles FitzRoy, sailing out to take up his post as Governor General of New South Wales.

If nothing else, the New Zealand posting had given FitzRoy the opportunity to revisit the oceans on a long sea voyage. He must have made an impossible passenger, pacing the deck of the *David Malcolm* with a keen eye on the ship, the performance of the crew, and the weather. One evening, the ship anchored in the harbour of Mercy at the westernmost end of the Magellan Strait, a particularly desolate and savage shore at the tip of South America. The night appeared balmy and calm, but no one knew this coast and its weather better than FitzRoy, and he was not deceived. His instruments told him there was a storm brewing. He related such to the Captain, advising that the second anchor be shipped, allowing the ship to veer into the changing weather. The Captain, who 'neither had nor cared for a barometer', would have none of it, and retired for the night. FitzRoy stayed up, keeping his eye on his barometer, and when, approaching midnight, it fell alarmingly, he went on deck. The officer of the watch, young and appreciative of FitzRoy's experience, disobeyed his Captain and heeded Robert's advice. He took a risk, for countermanding orders amounted to mutiny and, as the night continued calm and clear, he must have been wondering what on earth he had done, believing FitzRoy.

> But about two o'clock . . . a roar was heard on the western heights and in a few minutes that ship was nearer on her beam ends than at any time when under sail . . . had that ship been taken unprepared, not a soul would have survived in human probability; only God's providence could have rescued any one in so desolate, wild and savage a country.

Back home and jobless again, with the FitzRoy finances dangerously reduced, he took up his pen and looked for somewhere to occupy it. It was around this time that, over in Washington, Matthew Maury was studiously collecting and collating facts and figures on the winds of the Atlantic provided by returning sea captains. When he published his wind and current charts for the North Atlantic, FitzRoy must have been one of the volume's keenest European readers. After all, he had made his own observations of the winds and currents in the dangerous seas around South America. This was the spur he needed and he set to work, producing his own companion volume, his *Sailing Directions for South America*.

Soon after publication, the Admiralty came up with another post for him, as Superintendent of Woolwich Dockyard, where, combining his scientific and technical astuteness with his undoubted abilities as a sea captain, he was given command of the *Arrogant*. This ship was as far removed from the *Beagle* as it possibly could be – she was not a small sailing barque, but a screw-driven steam-powered warship. Neither was he to embark on a long voyage of exploration – he was appointed solely to supervise its fitting out and trials. For once the weather did not occupy him quite so urgently, chugging up and down between Woolwich, Portsmouth and Portugal in his steamship.

In 1850, the trials of the *Arrogant* completed, FitzRoy decided to retire from active naval service at the age of 45, citing poor health and personal problems as his reasons. His old friends, Darwin and Beaufort, wishing his contribution to science to be officially recognised, backed his election to the Royal Society, and in 1851 he became a Fellow. The honour might well have given a boost to his ego but was of little financial help. Among the personal problems precipitating his early retirement might well have been the health of his wife. She died in 1852, leaving him with four children – the eldest a girl of 13 – outstanding debts and a sorry-looking future.

However, stirred by Maury, events in Brussels, and FitzRoy's proven aptitude for meteorology, that situation was about to change. In the autumn of 1853 the first of a series of communications from Colonel Edward Sabine came to his door.

EIGHT

The Birth of the Met Office

Robert FitzRoy must have been surprised and flattered when the then Secretary of the Royal Society, Edward Sabine, began a personal correspondence with him. FitzRoy, himself a Fellow of the Royal Society, naval surveyor of note, and a man with a passionate commitment to improving safety at sea, had probably made no secret of an interest in the proceedings at Brussels, and had quite probably spoken to other Fellows, perhaps even Sabine himself, regarding the outcome and the setting-up of a Meteorological Department. But he could not have expected the flow from Sabine of news of developments, copies of letters from luminaries such as Beaufort, Quetelet, Maury and others, nor the veiled suggestions that he, FitzRoy, himself, might have some ideas on how to organise a Meteorological Office and recommendations as to the qualifications of the man to lead it.

He was not slow to take up on Sabine's hints. On the morning of 5 November 1853, after receiving the latest missive from Sabine, FitzRoy settled at his desk and started to compose the draft of a memorandum recommending the selection of a 'zealous, trustworthy, efficient officer with good eyesight, who is able and willing to devote himself for years to come to the great object you have in view'. As far as we know, this memorandum was never sent. Although he was not known for holding back, perhaps he felt it was too risky to be suggesting himself before any official announcement regarding an office had been made, or perhaps it was because the situation was still so fluid that he could not decide to whom to send it. Whatever, he stowed the document away, keeping it safe while, like many others, he watched for further developments.

Sabine's other concern was the matter of the supply of suitable instruments. Vast quantities of barometers, thermometers and hygrometers would be required to be

distributed to all the vessels it was envisaged woul
part in the exercise. What is more, in order to pi
most useful readings, the instruments would hₐ
accurate and regulated against a common standard. ..ₑw
Observatory was regarded as the chief testing and regulating
centre for astronomical and meteorological instruments,
making it the ideal testing centre, and Sabine was General
Secretary of the British Association, which owned and ran the
observatory. So, without waiting for any government go-ahead,
Sabine had steered Kew into preparing specifications for the
design and manufacture of suitable instruments.

Sabine had not been the only one eager to begin planning
the supply of instruments. Henry James, now stationed with
the Ordnance Survey in Edinburgh, was one step ahead of
him. Not content with preparing specifications, James had
already been holding discussions with leading instrument-
maker, Patrick Adie of Edinburgh, about the design of a marine
barometer. He had also, as he had promised Beechey he
would, been obtaining expert backing for his proposed land
conference, and now had written support from meteorologists,
including Burgoyne, Quetelet and other Brussels delegates,
plus Maury himself. By December 1853 he was concerned that
nothing in his barrage of correspondence to Beechey had
elicited a reply. So he sent him copies of the letters of support
for the land conference, once again pressing him to act and
adding that 'If I do not see by the papers that you were alive, I
might have fancied you defunct.'

A harassed Beechey hurriedly passed onto Thomas Farrer,
Secretary to the Marine Department, all he had received from
James. Do what you want with this, he told Farrer, but he
recommended that the matter of the land conference should
be referred to the Royal Society (its known opponents) for
advice – and that no matter what, nothing should be allowed
to interfere with the arrangements for sea observations agreed
in Brussels. Beechey's lack of support was a blow to Henry
James. It added to the rebuff he had already suffered from
Professor Airy, who, after hearing from Cardwell that there
would be no salary for the post, had declined, after all, the
supervisory role in the new Meteorological Department.
Effectively, James's plans were scuppered. Sabine in the
opposing camp had come out victor – and the way was now
open for his protégé, FitzRoy.

On Sabine's recommendation, Naval Hydrographer John Washington suggested that FitzRoy should be proposed for the post to Sir James Graham, First Lord of the Admiralty, and wrote to FitzRoy in January 1854, offering a salary of £500 per year in addition to his naval half pay, although he would have to manage for the first year without a clerk, draughtsman or accommodation. FitzRoy did not take long to consider. In February he evidently assumed the post was his. In a memo to Lord Wrottesley (which he copied to Colonel Sabine) he wrote:

> It would not be too difficult to carry on without a draughtsman, but time would be lost. The more I think about the subject, the more interested I feel in it – and so I shall forthwith prepare myself for regular work – by going to a convenient house – where I shall have air, room and light – and shall be able to work at home as well as in other places.

No matter how prepared FitzRoy might have been, there still remained the hurdle of persuading an Admiralty on the brink of a Middle Eastern war that it had time to devote to setting up a tiny new department as an attachment to the Hydrographer's Office. Fortunately, help was at hand in the form of James Heywood, another Fellow of the Royal Society and former pre-Brussels deputation member, who was also a Commons MP. On 6 February, Heywood asked in the House 'whether it was probable that an office would be established to co-operate with Captain Maury and the American Government in oceanic and other scientific observations' and Sir James Graham replied that 'in consequence of Captain Beechey's report of the Brussels Conference it was the intention to appoint an officer to whom the observations made both on board merchant ships and Queen's ships would be referred. A vote for this purpose would be taken in the Navy estimate.' Parliamentary permission was thereby granted for FitzRoy's post to exist – though not, as yet, FitzRoy's official appointment to it.

And so the fledgling Meteorological Department struggled into existence, as a one-man adjunct to the Hydrographer's Office. It had not been an easy birth, and its early days were to be equally fraught, as Beechey and FitzRoy came immediately into confrontation. Beechey, possibly because he was aware of the Admiralty's sponsorship of FitzRoy, had in January prepared his own counter-proposal that the Department of

Science – another branch of the Board of Trade – be extended to cover the meteorological role with the appointment of a head assisted by four sappers and miners from the Royal Engineers who were already conversant with the collecting and collating of meteorological statistics. But Heywood's parliamentary question and the reply from the First Lord of the Admiralty effectively blocked that course of action and Beechey's Marine Department was left out of the reckoning.

That is probably where it would have stayed, had it not been for events in distant Crimea. The Hydrographer's Office still required funding for its meteorological activity, small though it was, and in March estimates were included with the Navy's proposed parliamentary 'vote', or budget, for the coming year. However, on 28 March, Britain declared war on Russia. Now, with war a reality, the Admiralty needed to channel every penny of its vote into providing for the Navy's fighting strength. Beechey was not slow to seize his chance, and when the Board of Trade came to present its vote in June, it embodied an amount for a Meteorological Department under the auspices of the Board's Marine Department. The vote was approved.

So Beechey had won his new department, but he had also gained, much against his better judgement, FitzRoy – and FitzRoy, still sponsored and paid by the Admiralty as Meteorological Statist, had already started work. One of FitzRoy's first actions was to redesign the layout of the log that had been agreed upon in Brussels. His design was based on a format he knew from his own experience was practical to use at sea, and he had both Sabine's and Beaufort's backing for this. Beechey, though, was adamant that the log must stay as it was, and in this he was supported by Maury. Representatives, including Henry James, had spent too many hours thrashing out the format just for FitzRoy to change it at a whim. The argument descended into a battle of two intransigents, both with their own, differing, ideas of the rules under which the department should be run. The Board of Trade's solution to the impasse was one that was not best designed to ease Beechey's dyspepsia – it appealed again to the Royal Society, which had, of course, originally proposed FitzRoy. Beechey attempted to counter this with an invitation to Henry James's erstwhile ally, Professor Airy, to proffer his opinion.

Both the Royal Society and the Astronomer Royal, it appeared, wished to consider the question fully before furnishing a set of objectives for the department and no replies were immediately forthcoming from either. In the meantime, without any officially stated objectives for the Department, Beechey and FitzRoy were set to remain at loggerheads, with an office-less FitzRoy operating from the comparative safe distance of a room at his club, the Athenaeum, while Beechey instigated a policy of non-cooperation.

All the infighting and indecision, which were such a feature of Britain's struggle to implement the Brussels agreement, do not appear to have arisen to the same extent elsewhere. For Maury, on his return to America, it was basically business as usual with regards to the operation of his Depot of Charts. Ships were issued with the new format of logs, wind charts were dispatched across Europe as promised, but otherwise the job of distributing instruments, to both naval and marine shipping, continued as before, as did the analysis and collation of returning data.

In France, it was the Ministry of Marine that took on the responsibility for putting the Brussels procedures into operation, immediately enrolling all naval vessels and encouraging the voluntary participation of merchant shipping. Christoph Buys Ballot in Holland was appointed to create a Meteorological Institute, which was soon up and running in Utrecht, and Spain had lost no time in contacting James Glaisher to test instruments for its observers.

FitzRoy, too, had a firm idea of what needed to be done and, of his own accord and despite non-existent resources, he set to work identifying and enlisting naval and mercantile ships for observational duties, appointing agents around the country, investigating the supply of suitable instruments and writing instructions for taking observations and completing logs to his specification. Nothing he did, though, received official sanction from the Board of Trade – until the autumn, when Beechey was taken ill. FitzRoy leapt at the opportunity, and, in Beechey's absence, he submitted his plans plus his own design of log to Lord Stanley, President of the Board of Trade. Lord Stanley, unaware that FitzRoy's log did not quite fit the Brussels requirements, appended his approval. FitzRoy immediately distributed it to his agents, along with his pamphlet of instructions, purposely omitting the amendments Beechey had ordered. Beechey, convalescing in the Lake District, could only fume.

While the rumpus between Beechey and FitzRoy gathered momentum that November, a storm of even greater import broke out in the Crimea. This was 14 November 1854, that terrible morning of gales and rain and thunderous seas at Balaclava. When the storm finally dissipated, the British naval fleet lay decimated, and the French had suffered the loss of their mighty flagship, the *Henri IV*. The consequences to the Crimean campaign were to prove catastrophic. Shock waves reverberated into the hallowed calm of Whitehall and rocked the peace of the Paris Observatory.

Not that there had been much peace at the Paris Observatory. Meteorological observations and collations as per the Brussels Conference agreement were progressing without problem at the Ministry of Marine, but at the observatory, where separate scientific meteorological observations continued to be taken, the atmosphere was eruptive. The position of the staunch republican Arago as Director of the Observatory had been precarious since Napolean III had taken power in December 1852, and, when Arago died the following year, just one month after the Brussels Conference, his going provided the opportunity for a major reorganisation. The name was changed from the Paris Observatory to the Imperial Observatory, most of Arago's staff, republicans like himself, were swept out and Urbain Le Verrier, a lecturer in astronomy at the École Polytechnique and known Imperialist supporter, was chosen as Arago's successor. At the time of the Balaclava storm, Le Verrier was completing an investigation into the work of the observatory and of the various scientific projects undertaken there – prominent among them being the development of meteorological theory. The sinking of the *Henri IV* was to have a profound influence on Le Verrier's recommendations.

The storm that hit Balaclava had progressed across Europe before crashing upon the Black Sea, and it struck Le Verrier that, had a telegraph line been in existence to the Crimea, the fleets and armies could have been forewarned of its approach. As a result, into his final report went a passage advocating the use of the telegraph as a mechanism for broadcasting reports of the movement and severity of stormy weather. Whether it was because of the loss of the warship, or because of Napoleon's vision of France as a world leader in science, or both, the Imperial Decree that laid down the future role of the observatory gave special prominence to its meteorological function, endorsing Le Verrier's scheme for the collection of simultaneous observations by telegraph and envisaging an eventual system for the transmission of telegraphed storm warnings.

Le Verrier did not confine himself to future recommendations. He took action there and then, requesting from observatories all over Europe, including Greenwich, their readings of barometric pressure, wind speed and direction for the days immediately before and after the Balaclava storm, and his meteorological officer, Glaisher's French equivalent, Liais, was set to work on analysing and charting the results. The outcome was a revelation. Liais's charts, on which he had plotted lines connecting places of equal pressure, showed high and low pressure progressing across Europe in a series of high crests and low troughs much like the waves on the sea, and that the severity of the storm increased with the deepening, or lowering pressure, of the troughs. Liais had additionally identified that a series of kinks, or abnormalities, in the smooth curve of the isobars had accompanied the storm across Europe, and deduced that storms could be foretold on a weather map by recognising similar distortions. Le Verrier was excited by the findings, hoping that the pattern identified by Liais might be developed to form the basis of a formula for reliable storm prediction.

The idea of waves in the atmosphere was not new. James Forbes, back in 1832, in a report to the British Association, had implied that atmospheric tidal waves were accepted as fact:

> The great extent of country over which the accidental variations of the barometer take place, is one of their most striking features; and in a future and more advanced state of Meteorology we may be able to draw the most interesting and important conclusions from the great atmospheric tidal waves which are thus perpetually traversing oceans and continents.

Liais's conclusions appeared to confirm their existence, and with a more complete set of observations than had been available since the Mannheim Ephemerides of 1780. Though they created a murmur of interest in the world of meteorology, few scientists were enthralled by storm warnings and it caused little impact outside France at the time.

In England, Robert FitzRoy did not need extensive fact finding to convince him that the approach of a storm could be identified by watching the changes in barometric pressure. Years of experience at sea had already taught him that. To him, the way to avoid a repeat of the Balaclava debacle was simple – give ships the right instruments, instruct their captains to use them properly, and they would be able to spot for themselves an imminent storm and gauge its severity. However, from where he sat behind the doors of the Athenaeum, his voice was weak and could not be heard. The after-effects of the storm at Balaclava, though, did kick up enough of an uproar to reach the ears of authority. Weather, and storms in particular, shot up the political agenda, and the task of gathering meteorological data suddenly became urgent. All hesitation, delay and neglect of the infant Meteorological Department dissolved and gave way to wholehearted support. Preparing for his Christmas break, FitzRoy received a document that arrived under seal from the Admiralty. Dated 24 December 1854, it confirmed his appointment as head of the Meteorological Department, effective as from the previous August.

At last his position was official – Captain Robert FitzRoy, RN, had become head of Britain's first government meteorology department, with the title of Meteorological Statist. Although he, personally, was appointed by, and reported to, the Admiralty, his department came under Beechey's Marine Department of the Board of Trade, thereby giving FitzRoy two bosses. It was a situation that was to prove almost impossible to live with, especially when one of them was determined to be as obstructive as possible.

The first home of the Meteorological Department was at No. 2, Parliament Street, which stood in the heart of Westminster. In 1855, when FitzRoy moved in, Parliament Square would have been alive with builders, masons, carpenters and painters, all swarming over the new Palace of Westminster, only recently emerged from the rubble of the old on the banks of the Thames. The two Houses of Parliament – the Lords and the Commons – had moved in just a few months before, but it would be another seven years before the building was complete. The west side of Parliament Street was then a huddled row of cramped buildings that stretched from Parliament Square to Downing Street, and No. 2 was at the far end of the northern block, now the site of the Home Office building.

The Meteorological Department of the Marine Department of the Board of Trade, to give the office its mouthful of a title, had appointed to it a staff of four: William Pattrickson, draughtsman, clerks T.H. Babington and F.R. Townsend, and William Leaker, statistician. Originally Leaker was to have been FitzRoy's deputy, but he lasted only until February, recalled to the statistical department from whence he had come when it refused to release him for duty. It was some months before Leaker was replaced, so FitzRoy was already understaffed when, in February, the Board of Trade received the long awaited report from the Royal Society. It was signed by Sharpey, now Secretary of the Society, but it was Colonel Sabine who had been the foremost mover in its compilation.

Speedy though Sabine had been in projecting FitzRoy into his post, he had been far slower and more methodical in producing the specification of the job he expected him to do. The Brussels Conference's primary objective had been to bring about improvements in navigation through international cooperation in the collection of meteorological observations at sea. The office set up to do this, Sabine was at pains to emphasise, should have as its prime responsibilities the collation and digestion of a mass of meteorological observations to be collected by ships, and the supply and control of the instruments necessary for the task. The conference's secondary objective had been to make available the mass of data collected in order to promote meteorological science. The Royal Society, and Sabine in particular, grasped with greed at this unique opportunity to fill the gaping void in the scientists' supply of worldwide meteorological data. Nothing on this scale had ever been attempted before. It was the answer to Forbes's heartfelt plea in 1832 for 'that essential condition which it seems the fate of Meteorology to want – Co-operation'. Sabine's report overflowed with detail on exactly what should be measured, how it should be measured, and what scientific ends the measurements were intended to achieve.

It was obvious that Sabine and the gentlemen of the Royal Society had no comprehension of the enormity of the tasks they were demanding a ship's captain to perform. But FitzRoy had, and he must have despaired at the scientists' disregard for practicalities of operations in the face of a gale. The document purported to be 'advisory' and FitzRoy took it to be just that. He read through the document, carefully noted its contents and went his own way, following the objectives he had already constructed for himself.

At this stage in his career, FitzRoy was still very much the sea captain and not yet the meteorologist. His personal inclinations accorded with those of Maury and top of his objectives was the preparation of charts or maps for navigators. The whole exercise would be reliant on the willing cooperation of ships' captains. Although he states: 'It is one of the chief points of a seaman's duty to know where to find a fair wind and where to fall in with a favourable current,' he considered that captains would be best encouraged by the promise of reward. The reward was hardly likely to be monetary – FitzRoy knew better than to suggest it – although it seems fairly obvious that a purse would hold more attraction to the hard-bitten master of a merchantman than a big, bold, 'Grade A' marked against his name on some government list somewhere. (In point of fact, those who excelled were given the price of a first-class telescope, though whether that was how they actually chose to spend the money is not recorded.)

Even before his post had been officially confirmed, FitzRoy was concerned that the department should be seen to be showing a return for the public money invested in it. One way of doing this, he thought, was to issue practical sailing directions from existing weather records as soon as possible. His inner eye glinted at the recollection of the Admiralty's vast store of ships' logs, diaries and reports full of useful information, and he had wasted no time in badgering the Admiralty for access. In a frenzy of activity that mirrored that of Maury's a decade earlier, he had already raided the Admiralty's attics and amassed a mound of documents. All that was needed now was for the weather data buried in them to be extracted and recorded. From the moment he gained a staff, he set his clerks to work on the task, and soon they were elbow deep in dusty old papers.

FitzRoy's desk, though, remained uncluttered. He had already engaged his agents and initiated the process of issuing instruments, logs and instructions to mercantile ships. Now he had to wait until sheets were returned, which could be a long wait, since in those days voyages could last for months. The Navy was preoccupied with Crimean War operations in the Black Sea, leaving little opportunity to contribute to a mere meteorological exercise. So, with everything well in hand, he gleefully turned his attention to his prime priority – the design and preparation of suitable charts. He began with a copy of one of Maury's wind charts of the North Atlantic.

Back in 1853, when Maury had been attempting to drum up interest in Europe for his Brussels Conference, he had been generous with the distribution of his wind charts. Even Beechey had been moved to order 100 of them, and some of these were now in FitzRoy's hands. Copying the model laid down by Becher back in 1831, Maury had divided his chart of the North Atlantic into a grid of squares, each measuring 5° of latitude and longitude along its sides. He had distilled all observations that had been made by ships sailing through those coordinates into a series of figures within the square, laid out in a shape he termed a 'wind rose'. Useful though this was, the charts were cumbersome for ships' masters trying to make the series of calculations needed to plot a best line of sail. It was possible, of course, and many had done it to great advantage – and indeed there were far more intricate computations carried out on a ship several times a day in the cause of navigation. But why make life so complicated? thought FitzRoy. How much better to represent the figures in graphic form, so that their meaning could be read at a glance. He decided to combine four of Maury's squares into one of 10 miles each side, designed a diagram to place in each and called it the 'wind star'.

The 'star' was made up of a series of lines radiating from the centre point of the square, each line representing a direction of the wind. The length of the line indicated the proportion of readings showing the wind in that direction. Thus, if the wind was observed most often as west-south-west, then the west-south-west line would be the longest. Further lines were drawn, joining up the tips of all the directional lines, and the result was a rather lopsided star shape, from which it was easy to see immediately which were the prevailing winds. As an aid to the reliance a navigator could place on the star, FitzRoy added two circles centred on the middle. The radius of one indicated the total number of readings on which a star was based, and the second showed the proportion of those readings on which calms were observed. So a star with a long westerly point, one very large circle and another very small circle showed that this was a region with a reliable westerly wind. A star with two circles almost the same size, and both small, indicated that this was a known area of calms, avoided by most ships. Basing his work entirely on Maury's figures, he set to work constructing a wind star for each square on the chart.

If it seemed that FitzRoy was trying to upstage Maury, then he was not intending to be underhand about it. He quite simply wished to use his new-found power to ease the lot of sailors, though the thought of a little personal glory to help wipe out the disasters of recent years could not have been completely absent from his mind. In fact, Maury was generous with his praise of FitzRoy. Post-conference, Maury had been following closely the progress of FitzRoy's appointment, kept informed by the indefatigable Sabine, and in March of the previous year, when FitzRoy's application to Lord Wrottesley was hardly dry on the paper, Maury had written to Sabine, saying that he 'anticipates much from FitzRoy that will be useful to the world'.

When FitzRoy came into the post, it was natural that the two should exchange official correspondence, for Maury was FitzRoy's opposite number on the other side of the Atlantic. It turned out to be a meeting of like minds, and, with Maury ever generous with his explanations and ideas and FitzRoy with much to learn, their letters soon developed into lengthy meteorological discussions. Out of respect for FitzRoy's navigation skills and his knowledge of the South Atlantic, Maury invited the Englishman's contributions to the forthcoming seventh edition of his *Sailing Directions for the Atlantic*.

Official correspondence from the USA arrived via the Foreign Office to the Board of Trade and thence to Beechey, to be passed on to FitzRoy. Gradually, FitzRoy became aware of Beechey's nosing into his letters, and acting as censor. The reason was quite simple. Beechey reckoned that anything Maury had to say, he ought to say to him – after all, he had been the Brussels delegate and was FitzRoy's superior officer. Not only that, it became very obvious from the letters that FitzRoy was flexing his muscles and harbouring ideas of a scientific nature that were way above his station as Statist. Beechey began to block correspondence from the USA, and FitzRoy was forced to ask Maury to write directly to him at 2 Parliament Street. Only brief official notes were passed through Beechey, while a full and cordial correspondence went on behind his back. Unfortunately, Beechey caught wind of their subterfuge, and was furious, threatening to cut FitzRoy out of official correspondence altogether, much to FitzRoy's frustration. It was a further symptom of the escalating difficulties between the two, a conflict that not only threatened FitzRoy's ability to perform his function in the way he saw it, but could jeopardise the way others regarded his position and even lose him his job.

On the other side of the Atlantic, Maury was fighting American officialdom in a similar battle of his own.

Simultaneous Observations

The letter to Maury from the Secretary of State for the Navy was not unexpected, but the contents were beyond belief: 'it becomes my duty to inform you that from this date [17 September 1855] you are removed from the Active Service List and placed on the Retired List on leave-of-absence pay.' This from the same man who only seven months earlier had been effusive in his support of a proposal to the Senate to reward Maury with the huge sum of $25,000 dollars (not granted, it is true to say) for his work on wind charts. Rewarded or not, Lieutenant Maury could nevertheless have been forgiven for believing he was eminently due for promotion not removal.

Early in the year, he had appeared before a Navy Board. However, this was no promotion panel. Officially named the Retiring Board, and popularly known as the 'Plucking Board', it had been convened by an overstaffed Navy to weed out old and incompetent officers. Everyone, from passed-midshipman to captain, had been sent before the panel, and Maury, after the praise and congratulations heaped upon him at home and abroad, felt he had little fear of losing his active commission. It seemed, though, that meteorological advances counted for as little with the American naval hierarchy as they did with the British; and the medals, honours, gifts and knighthoods that came rolling in from foreign countries, far from being an aid to his cause, were a downright handicap. Many of his fellow officers, and that included most of the members of the panel, were jealous of his success and openly hostile. Maury complained in a private outburst to his friend the Bishop of Tennessee:

> I have been brought into official disgrace – for what? I am as ignorant as you. The thing has been done by a board of Navy Officers sitting in secret, and acting mischievously. I neither know what my offence is, nor who my accusers are . . .

To this feeling – and a feeling of displeasure by no means uncommon to the Old Commanders – that I, a Lieutenant, have dared to establish a reputation somewhat honourable in spite of them, I ascribe their finding. I have supposed that the value and merits of the officer were to be determined to some extent by the fruits of his labour . . .

Maury immediately began a campaign to have his case aired before a court of justice. He figured on attempting, too, to shame the Navy into reinstating him by enhancing his reputation in the maritime field. He still retained his post as Superintendent of the National Observatory, which left him with some respect in scientific circles, and he determined to play on it. Maury was a surveyor and a man with the ability to visualise on a grand scale. Instead of restricting maps to what you could see of the ocean where it met the coast, he asked himself, what if you took that one stage further and made a map of what you could not see? In other words, what about drawing a map of the ocean floor?

Excited by a new sounding device that could not only give accurate readings of deep water depths, but could also bring to the surface samples of the seabed, he extended his requests to mariners to include depth soundings of the ocean floor in their programme of observations. Using these readings, he was able to profile the bottom of the ocean in the way a geographer maps the contours of mountains and valleys on the land. In 1855 he published the first edition of his book *Physical Geography of the Sea*, a seminal work that marked the inception of modern oceanography.

Not everyone was full of praise. Joseph Henry and Alexander Bache were predictably swift in condemnation. They grudgingly recognised the strengths and breathtaking spread of his observations over the oceans, yet they found much to criticise in his conclusions. Sir John Herschel, who, it is true, rarely had a good word to say for American meteorologists, especially Maury, was equally deprecating. He remarked to Robert FitzRoy:

I have great confidence in Maury as a compiler of and director of navigation but as for his theories and speculations I must say that a more complete farrago of nonsense than his Physical Geography of the Sea it has seldom been my fate to meet with. There is hardly a page without some gross misconception.

These criticisms, repeated by various scientists, were to some extent merited, for Maury had indeed splattered his book with hasty conclusions from inadequate sources. Had he not done so, had he merely presented his facts, then creaky theory would not have undermined the monumental achievement of the physical mapping. Literally millions of separate observations were collated and analysed to provide wind charts, shipping routes, sea temperature charts, maps of ocean currents and ocean depths, and also a chapter dedicated to the seabed of the Atlantic.

It was Maury's good fortune that publication occurred just at the moment when the feasibility of a transatlantic telegraph cable was being discussed, and who better to turn to for advice than Maury? With his ocean maps, he was able to identify and pinpoint the 'transatlantic plateau' that would make the laying of the cable possible. In return, the cable was to make his name.

FitzRoy meanwhile, suffering under Beechey, found in Maury a sympathiser to his cause. It was to Maury that he complained at length about his lack of staff – by autumn of that first year the office was swamped by a sea of paper, and the clerks, still only three, struggled hard to make headway against the flow. Maury, from his amply staffed premises, commiserated. It was Maury who was entrusted with an early copy of FitzRoy's wind-star charts, and he duly gushed with praise. And FitzRoy, confident of approval, sent Maury his new improved version of the ship's log, to record weather observations at sea. He had always considered the current British version cumbersome, and watching his staff's contorted efforts to prise data from it had no doubt confirmed his opinion. He had been recording his own weather observations since 1820 and completing official returns since he first commanded the *Beagle*, so he felt he could do much better.

The criticism, when it came, must have shocked. Maury thought the log too complicated, and castigated FitzRoy for straying from the agreed Brussels format. He warned him that, in asking for too much information from ships' masters without obvious reason, he risked losing their willing cooperation. Had FitzRoy applied more restraint and less haste, perhaps he would have thought to consider Maury's own feelings – for it was Maury after all who had designed and pushed through the original Brussels format.

Maury, though, had another agenda. The vexed question of a conference on land-based observations had refused to lie down, and, now that the maritime system was well on the way to becoming established, Maury saw this as the next development in his meteorological thrust – and another front on which to fight for his naval reinstatement. He cast an envious eye over the Smithsonian's burgeoning meteorological project.

Joseph Henry's band of land observers was rapidly expanding. Observations were beginning to come in from outside the USA, from places like Canada and Mexico, and the portion of the institution's budget spent on the meteorological project was growing year on year. The observations arrived on monthly returns, in much the same way that Maury and FitzRoy received their data in bulk from ships' logs at the end of a voyage. Thanks to Samuel Morse and his dot-and-dash code, the commercial potential of the telegraph had been realised, and telegraph lines were snaking their way across the huge expanses of the North American continent, linking remote areas once reachable only by days of rough country travelling. Joseph Henry, as the man who had pioneered the electric telegraph, was not slow to spot the implications of this new, instant, means of communication. As early as 1847, in the Smithsonian Annual Report, he was envisaging the time when the eastern states would be able to benefit from telegraphed warnings of storms approaching from the west. By 1849 he was already in discussion with the telegraph companies to persuade them to take his instruments at their various stations and telegraph the readings, daily, back to him at the Smithsonian. Now this network had become a reality, with increasing numbers of stations participating in the scheme. Henry, though, was as implacable in his rivalry as he ever was. There would be no hope of reconciliation there.

European support for Maury's land conference plan promised to be much more forthcoming. In Britain, the Royal Society, too, was of the opinion that the time was ripe to pursue an international land-based objective. The society's consideration of Maury's initiative rested solely on its scientific merits without any of the personal antagonism that swayed Joseph Henry, and Maury's international standing was likely to act to his advantage. There remained the stumbling block of Beechey. Still reliant on Admiralty funding, Beechey was resolutely set against anything land based. So, needing to court his favour, Maury plumped for Beechey's side in his battle against FitzRoy in the argument over ships' logs.

He was to discover that to turn Beechey from his course was well-nigh impossible. FitzRoy could have told him that – and probably did, for he found himself continuously baulked by Beechey. In the summer of 1855 came a major altercation over the Meteorological Department's first official report. Full of enthusiasm, FitzRoy had rushed to draft a document that was a typical FitzRoy production – rambling and subjective, with little to say on a department that, at this stage, there was little to be said about. Had he kept to the reasons for the existence of the department, stating its objectives and laying down how these were to be achieved, and left it there, all might well have been fine. But he insisted on launching into his personal agenda, furnishing his view of the Balaclava storm, which he claimed could have been predicted, and an account of his wind-star charts and their undoubted benefits to commerce and the well-being of seamen. Beechey duly blocked its publication.

Was FitzRoy surprised? He certainly should not have been. No way would Beechey be seen to endorse a report showing that his Meteorological Department had aims far exceeding the Royal Society's brief. There was nothing in that brief about producing wind charts to rival the already successful Maury charts, and as for daring to suggest that a storm could be predicted – and warned about – this was presumptuous in the extreme. Irritated by FitzRoy's audacity, Beechey immediately set to and drafted his own report, emphasising the Admiralty's contribution to the department in order to ensure their Lordships' continued funding. He made the mistake of assigning FitzRoy's name to the document. FitzRoy flew into a rage and crossed his name off the draft. Fizzing with fury, Beechey rushed off a second draft that was little short of a personal tirade against FitzRoy, castigating him for his attitude and his departures from principles agreed with the Royal Society. Fortunately for FitzRoy, it seems that reason prevailed at a higher level and neither of Beechey's attempts made it into print. FitzRoy, though, was determined not to be thwarted. From his own thinly lined pocket, he financed a private print of his original report, which he then sent out to a selected few

– Maury included. Much to Beechey's glee, no doubt, it appears that this unofficial document of FitzRoy self-praise created not a ripple of interest anywhere.

A relationship, already abrasive, was barbed all the more by this exchange, leaving a frustrated FitzRoy impotent against his battleaxe of a boss. Glimmers of light relief must have been few, but one such broke through one June day in the summer of the following year, 1856. It arrived by mail coach from an address in Yorkshire. In his position, FitzRoy received various appeals and letters from an assortment of people, all wanting to expound their own theories, usually wild and far-fetched, on all manner of vaguely weather-oriented topics. Beechey was quite happy to shuffle these cranks off in FitzRoy's direction. Most were dull flotsam, but in their midst the occasional gem sparkled, and such was the case with this. An innocent enough epistle, it came from a sea captain named Henry Clifton Sorby, of Sheffield, and it was a soundly constructed paper on Sorby's theory of the motion of waves.

The paper triggered an immediate reaction in FitzRoy. First, because he was so accustomed to acrimonious exchanges with Beechey, he must have found it a pleasure to be addressed politely and intelligently by a fellow seaman. Secondly, Sorby's views were entirely concordant with those of his own, which he had put in print as an appendix to his *Narrative of the Voyage of the Beagle*. But more than that, Sorby set him thinking along new lines. Scientists had claimed for years that air had the properties of a lighter version of water, that the atmosphere moved over the globe in a series of waves and tides and that the 'lows' and 'highs' of atmospheric pressure could be compared to the troughs and crests of waves in the ocean. But what if the atmosphere behaved exactly like the sea? What if the waves gave clues to impending weather, what if a storm was heralded hours in advance by an increasing swell in the sky? What if, in Sorby's theory of the motion of the waves, lay the key to foretelling the weather? He took up his pen and in a tone of confidentiality admitted to Sorby that 'your paper on the motion of waves is very interesting to me – and has suggested consideration in connection with movements of the atmosphere – as well as the Ocean – which when matured I hope to submit to you'.

From FitzRoy's tone it is evident that he felt he alone was on the verge of a discovery that could revolutionise all thinking on atmospheric theory. All through the latter part of 1856, while his pen was busy scratching columns of barometric readings, his mind was cogitating over circulating waves of atmosphere and how to prove their connection to storms. It was in transcribing statistical returns into graphical wind stars that the answer came to him. He needed charts of the atmosphere, showing the pressure at different places as a snapshot in time. Simultaneous observations were the key – ships' captains all over the world taking out their barometers and measuring the pressure, each at exactly the same time. And he, issuing barometers, logs and instructions in how to complete them, was in the ideal position to put the plan into action. It needed only a small addition to the instructions, a requirement that, as far as practicable, observations should be made at set times Greenwich mean time, time that they had to keep for their longitude reckonings. There was no way Beechey was going to wear anything like that, though. Any further change to his beloved logs would drive him apoplectic and, for the same reason, Maury would not be of help to him, either.

The notion of collecting observations at simultaneous times in separate places was not an original idea. Various attempts had been made in the past, notably with investigations into weather in connection with the earth's magnetic activity. Previous efforts, though, had always fallen down on the vast organisational effort involved. For this reason, the Brussels Conference in 1853 had been the ideal opportunity to raise the issue again. Beechey, like others, had officially backed the notion, although he did nothing to further it himself. However, after the Brussels Conference, the concept of collecting international simultaneous observations became enmeshed with the question of land observations, which, with Beechey's stubborn and continued blocking, faded from the international agenda.

But, while Britain dallied and FitzRoy confined himself to the sea, other nations persisted with their own systems of simultaneous land observations. In France, Le Verrier's ambitious plan for a telegraphed storm-warning service was beginning to take shape. While it was true that he was as yet not issuing warnings of any kind, he had been building a network on which they could be based. The French telegraph service was public rather than privately run, administered by the Administration des Lignes Télégraphique, and in 1856 Le Verrier succeeded in persuading the administration to his cause, with permission to use telegraph operators as meteorological observers, creating a telegraphic network much the same as Joseph Henry's.

Henry, as we have already seen, recognised the potential for using telegraphed intelligence of weather from the west as a harbinger of approaching storms for the east – with, given America's huge hinterland, a greater probability of accuracy than in France. It was a role he did not envisage for the Smithsonian itself, but rather thought should be the responsibility of a separate government department. With no such department in the offing, the storm-warning idea made no progress in America.

The French arrangement, though, was ideal – the telegraph operators could be controlled via a central administration, and their readings, taken as part of their everyday duties, could be relied upon to be regular and made at specified, simultaneous times. It was the kind of discipline that Colonel Reid had been able to call on from his teams of Royal Engineers' observers, and far preferable to a motley collection of volunteers, no matter how enthusiastic. Liais designed and distributed an easy-to-use barometer to twenty-five stations throughout the country, and by the end of the year they were taking three sets of observations per day, using the telegraph to send the early morning readings back to Liais at the observatory. As well as using the results for charts and analysis, Liais began publishing them in a Paris evening paper in 1857, and later that year a daily weather report was issued to foreign countries directly from the observatory. Beginning with just a single sheet, this *Bulletin météorologique international* would later expand and become one of the most respected international meteorological journals.

Maury meanwhile, snubbed by Joseph Henry and smarting from it, was looking for a way to break into Henry's monopoly of land observations. He hit upon the ranchers and farmers. The idea was not new to him and he had attempted to seed his proposal over the years without it taking root. Now had come the time for strong action. While one set of Americans had been garnering wealth by trading across the oceans, another set was harvesting the riches of America's vast prairies. Maury, of course, had his own roots among farming pioneers, and he knew how to appeal to these people. Instead of preaching science, Maury took the same line as he had with the sailors –

hinting at how, in return for their cooperation, he could provide the information to save them time and money.

> The atmosphere is a great basin which envelopes this globe, and every plant and animal that grows thereon is dependent for its well-being upon the laws which govern and control the 'wind in its circuits', and none more so than man, the lord of all. To study these laws, we must treat the atmosphere as a whole. We have now the sea made white with floating observatories . . . We want to see the land, therefore, spotted with co-labourers observing also, according to some uniform plan . . . I have addressed myself to the agricultural interests of the country, because they have the deepest stake in the fence.

He set out on a grand tour of agricultural societies in the south and west, delivering the same message. Volunteers flooded in, offering to make the observations of temperature, winds and rainfall, along with the accompanying condition and yields of crops. State legislatures voted funds for observatories and collection points. Maury's unofficial correspondence to FitzRoy would most probably have been full of effusive descriptions of his progress with this agricultural-observations programme, and FitzRoy's replies supportive. Quite suddenly, however, the letters from FitzRoy stopped. It is true that both men were entering a fraught and frantic time in their personal affairs, but even so, as sympathetic friends and energetic letter-writers, it seems strange each should deny himself the pleasure of expounding his own theories to a receptive and unbiased ear.

Maury's crusade was being fought against the US Navy and Congress in his struggle to overturn the verdict of the Plucking Board. In England, FitzRoy's battles were being fought with Beechey. Their feuds, instead of calming over time, had become increasingly acrimonious. But Beechey was not in good health – it was his enforced absence in 1854 that had allowed FitzRoy the chink he needed to start work. No doubt illness only served to make him all the more crotchety, but FitzRoy allowed him no quarter, riling against him at every opportunity. Animosity and constant aggression did not improve Beechey's health and, in November 1856, he succumbed to illness and died.

It is highly unlikely that FitzRoy shed many tears at his passing. FitzRoy would have hailed Beechey's death as a release and, more importantly, his chance for progression. He instantly applied for Beechey's former post and, on the face of it, should have been favourite for the appointment, for who else was better qualified, better placed and more instantly available to step into Beechey's shoes? However, the Crimea intervened in his life again, this time working against him. The end of the war brought to shore a flood of retiring naval officers, all searching for a job on land. War at sea had given them the opportunity to shine, an opportunity denied to FitzRoy bristling under Beechey's iron hand. It was one of these returning heroes, Captain Bartholomew Sulivan, who was appointed to replace Beechey at the Marine Department of the Board of Trade.

This was that same Bartholomew Sulivan who had been FitzRoy's one-time boy companion from the *Thetis* and had served as a promising junior officer on the *Beagle*.

The two had not lost contact over the intervening years. Indeed, only a few months before, while commanding his ship, the *Marlin*, Sulivan had been approached by FitzRoy to take on board a new design of ship's barometer and asked to try it out, along with the new meteorological logs. Sulivan had been happy to oblige, and corresponded with FitzRoy and the barometer's manufacturer, Patrick Adie, in some detail about the pitching on board ship that caused the barometer to break.

Sulivan had never lost his admiration for FitzRoy. Robert FitzRoy was the role model Sulivan had hero-worshipped, who had taken him under his wing and taught him to watch the winds, observe the weather and follow the stars. He respected him still, and the job he was attempting to perform. Now their roles were officially reversed, Sulivan would not feel comfortable lording it over his former commanding officer, and he had no desire so to do. His decision was to allow FitzRoy full rein at the Meteorological Office, reporting directly to the Board of Trade without referral to the Marine Department. Nominally, he was still FitzRoy's boss, but in practice FitzRoy was free to go his own way. It was an arrangement that suited FitzRoy – and a hefty consolation prize for his missed civil preferment arrived soon afterwards in the form of his promotion in rank to Rear Admiral. The time was propitious for him to launch the first of a series of initiatives that were to lead him from the back waters of mere statistics into the murky, shark-infested seas of the emergent science of meteorology. It was not the time, though, to reveal all to a would-be competitor on the other side of the Atlantic.

FitzRoy lived in fear of plagiarism. He knew his idea of simultaneous observations was not new, and he also knew of the achievements of Maury and Henry in America – or, rather, he had been told Maury's version of what was being achieved. Where FitzRoy's initiative differed was that, while Henry wanted to study storms on land and Maury wanted to gauge the effect of the weather on crops, FitzRoy's obsession was with the motion of atmospheric waves over the Atlantic. If he wrote to Maury, it would be difficult to hide his own plans. Maury's forte did not lie in original theoretical thinking, and the last thing FitzRoy needed then was for Maury to seize his idea and dash enthusiastically into the study of atmospheric waves. It was inevitable that, if he did, the world would laud the famous, well-respected American in place of the naval statistician upstart, FitzRoy. And FitzRoy desperately wanted to make his mark on his own account. Now suddenly, Beechey was gone and Sulivan had opened up a direct line straight to the top. He spent his Christmas perfecting his proposal and with the start of the new year he was ready to let it loose.

If you are going to make a splash, you might as well make a big one, and FitzRoy prepared to launch his proposal across several channels, all at once. First of all he drafted a memo to be sent to the President of the Board of Trade, Lord Stanley; the Vice President, Robert Lowe; and its Permanent Secretary, James Booth. In it he proposed his system for mapping a simultaneous picture of the atmosphere, describing how he would incorporate various states of wind and weather and project a series of observations onto a succession of charts to trace the atmospheric waves. Next, he prepared a letter for publication in *The Times* and the text of a Public Notice, inviting the public at large to send him their own readings. Judiciously, he also submitted a copy of everything to Thomas Farrer, Secretary at the Board of Trade.

The memo came into Farrer's hands on 6 January 1857. Astounded, he rushed to intercept the copy intended for the President and went in haste to Booth and Lowe to seek their views. Two days later, FitzRoy was devastated to learn that his memo had been withheld from the President. It was not the idea of simultaneous observations that upset Farrer, Booth and Lowe, nor were they pouring scorn on his idea of waves in the atmosphere. No, what they objected to was his plan of inviting all and sundry to send in information. Just how reliable did he think this would be? No scientist would ever take his results seriously.

FitzRoy fumed. He might as well still be fighting Beechey. After so much thought and preparation, he was not ready to let his bid for fame and glory simply lie down and die. He dashed off a reply to Robert Lowe. The rubbish, he retorted, could easily be eliminated from the public returns, leaving sensible, reliable data. Using Maury and his reputation as vindication, he said: 'Maury is now working out a similar problem in America. He expects to give notice of storms, along the coast, by telegraph – as there the hurricanes and most gales move simultaneously.'

In fact, this was not quite true. Maury was preoccupied with farmers and their crops. If anyone, it was Joseph Henry who had it in mind to utilise the telegraph for storm warnings. But quite probably it was the truth as FitzRoy had perceived it from Maury, and anyway it served its purpose. It enabled him to leave Lowe with a brief, enigmatic glimpse of the future – an expectation of his own to be able eventually to give notice of storms by telegraph in Britain. However, he submitted dutifully that, since it was the wish of everyone that he abandon his plans for gathering simultaneous observations, then he would, for the moment, put the subject by.

In reality FitzRoy had no intention of putting the subject by – quite the contrary. If he were not to be allowed to gather the new observations he needed, then he would just have to scout around for historical data that fitted his criteria. He resorted to the set of weather readings gathered throughout Europe at the time of the Balaclava storm. On his charts, FitzRoy did not use Liais's method of drawing isobars. He believed then, and continued to believe, that isobars drawn from insufficient numbers of barometer readings served only to hide rather than to reveal the subtle variations of pressure caused by wind speed and by land features such as mountains, lakes and forests. Instead, he used his own methods of charting, simply by spotting the charts with readings in the location where they had been taken.

However, no matter how he pored over his charts, they refused to yield to him any further enlightenment on his atmospheric-waves theory, and how it might apply to the nature of storms themselves. Though he was well read on the subject of storms – he had studied experts such as Redfield and Espy, learnt from Reid and heard what Maury had to say – still there loitered something he could not quite grasp, the missing link to take him from their theories to the practicalities of the charts before him. He needed to know more, and the person to turn to was the world's current expert on the subject, Henry Dové. Dové had published a book, *Das Gesetz der Stürme* (*The Law of Storms*) which FitzRoy was anxious to read, yet the only copies available were in the Prussian's native language, German. FitzRoy began to press for a translation into English.

Setback only seemed to spur him on. Simultaneous observations became his preoccupation, and, to some at least, it appeared that FitzRoy was neglecting his duty. Criticism was building. The Meteorological Office had been in existence for three years, instruments by the score were roaming the high seas in ships of all shapes and sizes, log after log was disappearing daily through the Office doors, yet very little was re-emerging. Where were all the results that had been promised? The truth was, FitzRoy's three clerks were struggling with the volume of work. Returns from mercantile ships alone were coming in at the rate of about one a week, each return containing an average of seven months' worth of figures. Add to that the naval returns hitting him with a rush now the Crimean War was over, plus the mountains of historical data still to digest, and it was all becoming too much to handle. FitzRoy thought it judicious to stem the criticism with the publication of his first independent official Meteorological Report.

It was on a May day in 1857 that he sat down with his pen and a blank sheet of paper and, with feeling, began: 'I am enabled to write with rather less limitation than hitherto . . .'. He launched into details of the department's work, and issued a plea, very reminiscent of Maury's a decade earlier, for more space and more people. And then, in a move he would not have dared risk under Beechey, he could not resist throwing his own pet project into the pot. He was submitting, he said, a plan to obtain 'simultaneous states of the atmosphere' from some land-based observation points, lighthouses, for example, in addition to the maritime returns. In justifying the diversion of official effort and time to his own purposes, he pointed out that the results might be useful, too, to scientists seeking to prove their theories. Despite his attempts to disguise his new venture within the terms of his brief, it was evident he was taking the first tentative steps on a route that would lead him way beyond the bounds laid down by the Royal Society. From now on, there would be no going back.

He rushed the report out into the hands of Parliament and the public in August, and considered his way now clear to start to gather simultaneous observations, maybe not with the big blast onslaught he had first hoped for, but with insignificant, insinuating little steps. He had already inveigled data from the Ballast Office in Dublin and the Northern Lighthouses Office in Edinburgh, both being about as far away as you could get from the prying eyes of the Board of Trade in London. Continental newspapers also proved fertile ground with their regular published weather reports that he could net into his collection. Becoming more daring, he approached Trinity House and then, via his agent in Liverpool, a certain Mr Hartnup, he managed to persuade Mr Cunard to take the forms on his yachts, with the promise of a fee of 5s for his trouble.

All this activity inevitably reached the notice of Lord Stanley, President of the Board of Trade. However, far from censure, the result was sanction: official permission to collect and collate from these new sources of observation. As soon as it came, FitzRoy was scribbling again, this time instructions to his draughtsman, Pattrickson, to begin work on drawing new charts, charts he termed his 'Synchronous Wind Charts', the new, improved version of his wind stars based on simultaneous observations. Wind charts were far from being FitzRoy's ultimate goal – proof of his theory of atmospheric waves was still where he was heading – but Lord Stanley's approval of his small start was to him a huge step forward, and he felt himself safely on his way. So safe, in fact, that he chose now to renew his correspondence with Maury.

It was not an ideal moment to choose to reinstate a relationship. Maury was facing a testing time, with all his troubles coming to a head at once. On the other hand, he probably needed all the friends he could get, since in America he appeared to be surrounded only by foes. He had eventually persuaded the farming community to his scheme for agricultural observation, and had set it into motion. America now had two systems of land observations, and it was not long before people began to question whether Maury's true purpose was to rival the Smithsonian. Maury was strong in his denials: 'Take notice now, that this plan of crop and weather reports is "my thunder"; and if you see some one in Washington running away with it, then, recollect, if you please, where the lightning came from.'

Yet the accusation stuck. It must be remembered that at this time, in 1857, the 'United States' of America were not quite so 'united' and the talk was of secession. Although Maury detested the whole idea of civil conflict and actively campaigned for conciliation, he was nevertheless a Southerner, living and working in the North, a situation that was becoming more and more uncomfortable by the month. Joseph Henry, on the other hand, was very much a man of the North and therefore held the advantage over Maury within Washington and New York circles.

Certainly, Joseph Henry was the darling of Washington. He had established the prestige of the Smithsonian Institution, and his meteorological work was highly respected, not only in America, but worldwide. He had also succeeded in bringing meteorology to the people. Within the imposing entrance hall of the institution, he had set up a large map of the USA and every day the telegraphed weather reports would be posted upon it, different coloured discs indicating the conditions at various locations, and arrows showing the direction of the wind. It was the world's first public weather map. People flocked around it, and it became the place for socialites to be seen. Tourists pointed excitedly at the weather back home, while regulars began to recognise that a morning soaking in Cincinatti frequently heralded an evening of rain for Washington. The *Washington Evening Star* was not slow to realise the weather's popular appeal. From May 1857, at about the same time as Liais's reports started to appear in the Paris press, it began to publish Henry's telegraphed reports from cities around the country.

As a young man, Henry had yearned for a starring role, and at last he had managed to steal the meteorological spotlight from Maury. Having reached the top, he realised how difficult it would be to stay there, and he was constantly looking over his shoulder, ready to defend his position against anyone who dared attempt to upstage him. One such was Samuel Morse. Henry always maintained that it was he, not Morse, who had invented the electric telegraph and he resented Morse claiming the glory for himself. However, while it was true that Henry had invented a form of electric telegraph, he had not been the only one to make a similar discovery. Furthermore, without Morse's acumen in recognising the commercial feasibility of a telegraph and his invention of a code to make this possible, a nationwide messaging system would not have become reality. Battle lines were drawn between Henry and Morse. Henry, naturally, had Bache and his coterie championing him, while Maury, not surprisingly, pitched in on the opposing side, using a tour of north-west agricultural societies to speak out in praise of Morse.

Much in the way that the storm over storm theory had escalated, this new slanging match degenerated into bitterness and foul play. Bache and Henry lashed out at Maury, castigating him for neglecting his duties at the observatory while running round the country chasing his own ends. He was unfit to run the national establishment empowered with regulating the country's timekeeping, they sneered, and should keep to the sea. Maury struck back. Bache's Coastal Survey ought to be run by a seaman, he said, and campaigned to have it moved to the Navy. Sense was brought to prevail only by the intervention of Chief Justice Taney, who pointed out to everyone involved that they were representatives of their respective institutions, not private individuals, and they should stop their squabbling before the reputations of all were wrecked beyond repair.

They made a tentative truce. Bache kept his Coastal Survey, Henry continued at the Smithsonian and Maury returned to his observatory – but not merely to observe. His flamboyant and engaging speeches countrywide had left the agricultural communities agog to see some positive results from his scheme. So he gathered together his evidence and put to Congress a reasoned proposal for federal funding for a national scheme. Congress threw it out.

On 27 November 1857 Maury was summoned to appear before the court of inquiry into the Plucking Board's sentence. In the two years since his dismissal, despite his entreaties and in face of his legal pleadings, he had not once been given any reasons for the board's decision. The proceedings had taken place in secrecy and no minutes were kept. The members of the board had operated safe in the knowledge that never would their deliberations be made public. Now Maury had forced them into a position where a case had to be made, in order for it to be answered. They could not claim jealousy, they could not advocate envy and they certainly could not put forward wanting to get their own back as a justifiable reason. The only real evidence they could produce was that Maury's disability (on account of his coach accident injury) made him unfit for duty. At the trial, Maury submitted to medical examination and was declared sound. Now all he could do was endure the nail-biting wait for the verdict. It was at this tense moment that FitzRoy had unwittingly chosen to write to Maury once more.

It could not have been an easy letter to write. How do you explain a gap of over a year in what had once been a prolific correspondence? FitzRoy's excuses were weak, but, fortunately for him, by January Maury's troubles were over. The President reappointed him to the Navy's active list, and promoted him to Commander. Magnanimous in victory, Maury appears to have been forgiving of the hiccough in the relationship with FitzRoy, and the letters began again. On FitzRoy's side, certainly, there was a change in style, for he was guarded, studiously avoiding any mention of atmospheric waves and his work on synchronous charts. Perhaps he still did not trust that Maury would not move in on his territory, sucking FitzRoy in with him on a surge of enthusiasm, and allowing the world to believe the theory was Maury's own. Maury, though, was as forthcoming as ever on the work he was engaged in, and no doubt this was exactly what FitzRoy wanted to find out. The resumption in correspondence carried an ulterior motive on FitzRoy's part. Just what this was would emerge some time later, as his plans began to unfold.

With his 'slowly slowly' approach to simultaneous observations now under way, the powers that be at the Board of Trade were no doubt hoping that FitzRoy would at last get down to his real task and apply himself to tackling the backlog of ships' returns to be tabulated. However, the detailed work of computation did not appeal to the Board of Trade's Meteorologic Statist, who by this time had, in his own opinion, risen above such mundane tasks. He was already looking elsewhere, focusing on the next item on his own agenda. He had turned to fishing.

T E N

The FitzRoy Barometer

Nothing much ever happened in the sleepy Cornish fishing village of St Ives, so it must have caused quite a stir when the carrier's cart came rumbling over the cobbles and Admiral FitzRoy's barometer was unloaded from its rear. Out of its packing, the barometer stood at 3ft high, a sturdy instrument of glass and polished wood. Along its middle, a thermometer sat behind a glass cover, while at its top an oblong window revealed the upper part of a mercury column, flanked by white plates covered in writing.

That evening, the talk over supper in cottage kitchens and inns would have been the newfangled barometer sent by Admiral FitzRoy. Fishermen were distrustful of the weather glass, and did not think much of its professed ability to predict the weather. Its wisdom tended to be accepted by town-dwellers, but farmers and fishermen knew it to be nonsense and much preferred their own weather lore. Handed down through generations, country wisdom had been proved with the evidence of their own eyes. The admiral's bright, shiny instrument did not appear to offer anything better. What few of the fishermen had seen, and therefore did not yet know, was the brand new legend attached to the new barometer.

FitzRoy had spent many hours closeted with the manufacturer designing and perfecting a barometer that could warn the fishermen of approaching storms. The important change in emphasis, and the idea that FitzRoy wanted to convey, was that the pressure reading was an indication not of the weather at the present time, but of the weather to come, and for that to be best gauged one reading was not sufficient.

It is not from the point at which the mercury may stand that we are alone to form a judgement of the state of the weather, but from its *rising* or *falling*, and from the movements of immediately preceding days as well as hours, keeping in mind the effects of change of *direction*, and dryness, or moisture, as well as alteration of force or strength of wind.

FitzRoy had sent hundreds of barometers and other instruments out in ships all over the oceans, where seamen spent hours taking readings and entering them into logs. The logs were then brought back to shore and stored for later transcription and analysis, hopefully to benefit subsequent voyagers, perhaps years in the future. But what a waste just to note figures in a log. Why not let the seamen use the figures they were so assiduously compiling for their own immediate reward? With the addition of a small book of instruction, telling them not just how to record the figures, but how to interpret them as well, inexperienced seamen could learn to use science as well as lore to foretell the weather. The weather glass, properly used, could guide them to make better use of winds and perhaps avoid disastrous storms, like the one FitzRoy had himself fallen foul of on his first command of the *Beagle*.

So he had set to, and distilled all his experience of the weather at sea, plus all the knowledge he had since acquired, into a series of 'rules', which he wrote down in a small book he called his *Barometer and Weather Guide*, published in April 1858. The 'rules' were written as a plain man's guide, and were augmented by those seamen's sayings that to FitzRoy rang true. Such as:

> Long foretold, long last,
> Short notice, soon past,

which neatly summarised FitzRoy's rules 15 and 46, where he says that the longer the time between the signs and the arrival of the change foretold by them, the longer the changed weather situation would last, and vice versa. And:

> When rise begins, after low,
> Squalls expect and clear blow,

which emphasised his warning in rule 22 that the most dangerous shifts of wind or most severe northerly gales happen soon after the barometer first rises from a very low point.

On the barometer itself, on each side of the mercury column, he replaced the former, meaningless, legend with a new form of wording:

RISE	FALL
for	for
N. Ely.	S. Wly.
NW–N–E	SE–S–W
DRY	WET
or	or
LESS	MORE
WIND	WIND
———	———
Except	Except
wet from	wet from
N. Ed.	N. Ed.

For its time, the guide was unique. No one before had written anything that sounded quite so authoritative, yet was quite so easily understandable by the layman. 'A Capital little Weather Guide', Maury praised it, although predictably it found little favour with other meteorologists, who considered its mixture of unproven statements and homely folklore just too unscientific to warrant merit. However, his intended audience loved it, and that, for now, was all that mattered – or, almost all. There remained a section of the seagoing population who were totally neglected, and FitzRoy's next mission was to remedy that omission.

Then, just as today, it was always the big disasters that were newsworthy. When Britain's rocky shores claimed as victim yet another unlucky merchant vessel, newspapers proclaimed the horrendous loss of life and the sacrifice of cargoes valued in thousands of pounds. The Balaclava storm hundreds of miles away, with its disastrous tactical consequences, let loose cries of why did it happen, and how could it be prevented. Yet the greatest loss of British life at sea came not in these headline-hitting disasters, but regularly and unheralded to the hundreds of small-scale inshore fishermen who perished annually in the course of their trade. To the government, these tragedies existed only in figures from the Department of Wrecks, and in the consciousness of the public at large they featured hardly at all.

Large ships were now leaving port armed with every aid science could provide – barometers, thermometers, wind charts and FitzRoy's own weather guides. But there was nothing to help the fisherman. His biggest enemy was the weather, or rather a sudden, unforeseen change in the weather, when a storm could appear out of nowhere, leaving a small vessel helplessly at its mercy. FitzRoy's notion was to introduce a simple-to-read 'fisheries' barometer, and an even simpler weather guide to help in its interpretation.

As a principle it was highly laudable, but it fell down on one criterion, and that was money. The struggling fisherman could in no way afford the price of a barometer. Small inshore boats did not fit into the scheme of maritime meteorological observation that was the Meteorological Department's remit, so under current regulations FitzRoy could not provide instruments free of charge. Even if he did find some way around the financial problem, how would a skipper manage to take readings while fighting the elements and handling his nets? And then there was the delicate nature of the instrument itself. Despite the best design work that had gone into marine barometers by top instrument-makers and with help from FitzRoy himself, numbers of instruments were still broken at sea in the relative calm and order of top naval vessels. They were not nearly robust enough to withstand the conditions on board a battered fishing boat.

FitzRoy's solution was to set up the instruments on land, at fishing ports or 'fisheries'. There, a responsible person would take regular readings and offer weather advice to skippers before they set out. He thought that if he started small, by recommending just those fisheries that formed the 'top ten' of the casualties league, he might be able to obtain funding, on humanitarian grounds, to provide instruments free of charge. Convinced of its justification, he sent his proposal to Lord Stanley, President of the Board of Trade. Unbelievably, the reply was positive. Or rather, it came couched in the Civil Service terminology of 'agreement in principle' to the general idea of issuing barometers to fisheries, which FitzRoy chose to interpret as positive, giving him carte blanche to do as he pleased. He negotiated with manufacturers Negretti and Zambra to develop a suitable instrument based on his requirements, and in June 1858 the first barometers were sent to nine fisheries in Scotland and one to St Ives in Cornwall, plus, very soon afterwards, an additional issue, which went off to Filey on the Yorkshire coast. The citizens of Filey were blessed with a man of the church, a certain Revd Jackson, who cared as diligently for their bodies as he did for their souls. Way back in January, at the very moment when the Lord President was considering FitzRoy's case for fisheries' barometers, the reverend had applied personally to Lord Stanley for a barometer for Filey. His timing was unbelievably opportune.

The effectiveness of the fisheries barometer was not measured in figures – the loss of a fisherman's life did not feature in official returns – but if popularity is an indication of success, then it succeeded way beyond expectations. Such was the barometer's reputation, perhaps as much for the comfort of its presence as in its ability to save lives, that ports from other parts of Britain began to lobby for an instrument for themselves. FitzRoy attempted to grant as many of these as he felt he could. Sliding in the odd extra here and there seems to have gone uncommented, but come January 1859, when he boldly put through to accounts a requisition for twenty barometers and manuals, he once more brought upon himself the wrath of Thomas Farrer. FitzRoy fumed, but he was pitching his self-righteous impetuosity against the barricade of Civil Service bureaucracy – and was learning the hard way what an impenetrable barrier that could be. As far as the Board of Trade accountant, H.R. Williams, was concerned, agreement in principle did not mean permission to proceed, and FitzRoy was not just exceeding his budget, but acting beyond his authority. FitzRoy was soon in contention with Williams again, and this time about the costs of using independent carriers to transport barometers to the fisheries rather than routing them via the Coastguard division. FitzRoy retorted that he could hardly ask the Coastguards to deliver to the fisheries. Coastguards labelled all fishermen as either smugglers or informants and trusted none of them – and what is more, they did not take too kindly to the fishermen's dirty boots. In other words, the accountant knew nothing of the real world and should stick to his figures.

Although fending off the Board of Trade bureaucrats was becoming increasingly irritating for FitzRoy, the principle of supplying fisheries with barometers had been established. Some instruments were getting out there and would continue to do so. Meanwhile, work was continuing apace on his simultaneous observations. With an increase in staff and extra accommodation having been granted, FitzRoy felt free to allocate the task of constructing charts of these observations to one of his clerks, Thomas Henry Babington. Babington was to 'give a succession of synoptic views of a certain area of atmosphere from a point of sight above the lower strata, which are our winds (when in motion laterally)', and to show on those views the temperature, pressure and moisture, all taken at a time that approximates to the nearest quarter of an hour GMT. By such means, said FitzRoy, 'we may elicit the nature and character of successive graduations and places of change in direction and strength of wind, varying amounts of rain etc.'.

FitzRoy's use of the word 'synoptic' (from the Greek *sunopsis* meaning 'view together') in his March 1858 instructions to Babington appears to be its first application in meteorology. 'Synoptic' was commonly applied to the New Testament Gospels of Matthew, Mark and Luke, who each told the same story, the story of the life of Christ, from his own point of view. As a student of the Bible, FitzRoy would have been familiar with this usage, so it would come naturally to him, 'synoptic view' being much less of a mouthful than 'view of simultaneous observations'. The term would eventually be adopted as a standard, and the synoptic chart, the chart of synchronised observations, was to become one of the major tools of weather forecasting. In fact, FitzRoy described his synoptic charts as being 'as if an eye in space looked down on the *whole* Atlantic *at one time* and afterwards took similar views (much more extensive than '*bird's eye*') at regular intervals of hours and days, so as to obtain sequences of synoptic conditions' – a visionary description of today's weather satellites, 100 years before the first one was launched.

Babington's first synoptic charts were to be retrospective, drawn from figures already in the department. The period chosen was from October 1856, known as the Great Depression, to March 1857, and Babington's objective was to draw two charts for every day, at 9 a.m. and 3 p.m. GMT, using the measurements that had been collected for the preparation of synchronous wind charts. By the time FitzRoy wrote his official report for 1858, he had already concluded that the results were identifying atmospheric wave lines and evidence of a continuous alternation of great polar and equatorial currents of air. By stating in public his experiment and its conclusions, FitzRoy was openly revealing his department's new direction, and his own intention of crossing the divide from statist to scientist.

Even with his head filled with the future, FitzRoy still had the day-to-day matters of his office to attend to, one of which landed with a thud on his desk on the morning of 14 October 1858. The flow of ideas and inventions from cranks and hopefuls, the source that had produced the letter on wave theory from Captain Sorby, had never dried up. This packet was another of them. It contained sheets of detailed drawings and specifications that showed the hand of a skilled design engineer. On closer inspection, the invention it depicted appeared novel, and FitzRoy would immediately have grasped its potential. The instrument was called a Hand Heliostat, and its purpose was to use the sun's rays to flash signals across a wide distance. Ship-to-ship or ship-to-shore communication was an obvious application, and no doubt it was because of FitzRoy's connections with the Admiralty rather than the Meteorological Office that he had been approached. Excitement at the idea probably flooded through him, until suddenly it evaporated into scorn – at the point where he read the name of the inventor.

FitzRoy was not acquainted with the man in person but he was well acquainted with the name, for Francis Galton, of no mean renown in his own right, was the cousin of Charles Darwin.

Sun and Wind

The route to the top of Scafell, the highest mountain in the English Lake District, is a steep and steady ascent over grey slate and granite. Often the climber's only reward for his effort is a tantalising glimpse of a view through swirling mists of cloud. Sometimes, though, the skies clear, revealing, to the west, the Irish Sea, and northwards the curve of the Solway Firth with the hills of Galloway on its Scottish bank, while to the south lies the crinkled mass of the Langdales and, beyond them, Wales, with Snowdonia a smudge on the far horizon.

In 1841, when the young Francis Galton made that climb and reached the summit, it was not the peaks and lakes that caught his fancy, but a small group of men surrounded by charts and an assemblage of instruments. As he stared where they were looking, towards distant Snowdon, he would have seen a sudden glint of light that flashed again and again, quick then slow, a Morse coded message picked out in shafts of sunlight – a solar telegraph. A man on Scafell would have replied using a contraption of angled mirrors and shutters, called a heliostat. The men were from the Ordnance Survey, and they were experimenting with the long-distance, instant exchange of information their new heliostat afforded.

Francis's father, Samuel Tertius Galton, was himself a lover of mountains who liked nothing better than to wander the hills clutching his portable barometer. His passion was heights – not the climb nor the vistas but the measurement of them. Blaise Pascal, 200 years before, had proved that air pressure diminishes with altitude and Samuel Galton would appear to have been in complete accord with J.D. Forbes, who said:

> Of all the problems in Meteorology, few appear to me to be so intrinsically beautiful as that suggested by the fertile genius of Pascal – the application of the barometer to the measurement

of heights. It should, I think, be an object of ambition to bring this elegant method to the utmost degree of perfection of which it is susceptible.

Samuel Galton was descended from the notable Quaker family, the Barclays, and he, too, was a stalwart Quaker who could be strict and stubborn when he chose. A successful businessman, Samuel was painstaking and practical, not noted for flights of fancy or imagination. Exactly the opposite was Erasmus Darwin, Francis Galton's grandfather on his mother's side – an enigmatic character, as eccentric as Samuel Galton was conventional. From his work as a physician came a fascination for the functions of the human body, how and why they had been designed as they were. He dreamt up theories, some wildly erratic, some not so wide of the mark, though none of them did he submit to proper analysis. But then, detailed analysis was not to his taste. Where Samuel Galton's delight was to delve deep into statistics and mathematical proof, Erasmus's talents lay in imagination and invention. This was the man who, faced with a river separating his garden from his orchard, had designed a mechanical ferry to get himself across. He was also the man who had invented a wind-powered mill to grind dyes for the pottery works of his friend Samuel Wedgewood, and to whom another crony, Thomas Watt, had brought his plans for improvements to the steam engine for discussion and advice. The passion he invested in his science overflowed into poetry and if he did write of his scientific discoveries, he did so in verse, much to the irritation of his grandsons Francis Galton and Charles Darwin. Erasmus had died some years before they were born, yet he served as an inspiration to them both.

Francis's holiday in Keswick was a much-needed break in his studies to be a physician like his grandfather. Still only 19, he had already spent time as a house pupil at Birmingham's General Hospital, where as a 17-year-old with only a few weeks of rudimentary training he was performing operations – cutting off fingers and extracting teeth – not very satisfactorily it has to be said, certainly not for the poor patients. Fortunately for the safety and well-being of the citizens of Birmingham, he left after a year and went to King's College London to study anatomy and thence another year later to Cambridge, full of advice from his elder cousin Charles to the effect that careful observation rather than abstract reasoning

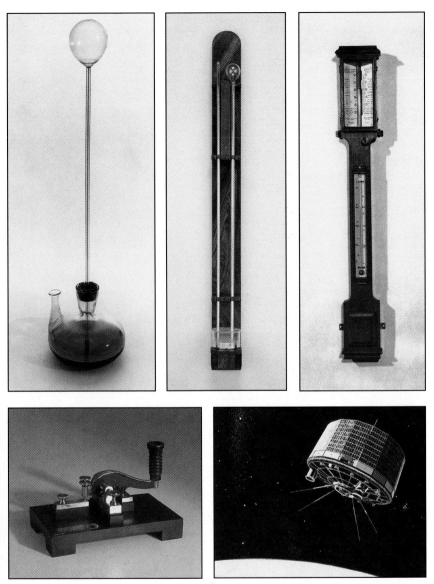

Top, left: Galileo's thermoscope. A reconstruction of Galileo's sixteenth-century experiment in the measurement of temperature. (*Science Museum/Science and Society Picture Library*) *Top, centre*: Torricelli's barometer. A replica of the 4ft tube inverted over an open dish of mercury, as Torricelli used in his original experiment in 1643. (*Science Museum/Science and Society Picture Library*) *Top, right*: 'Fisheries' barometer of a type distributed by Admiral FitzRoy from 1858. (©*National Maritime Museum, London*) *Above, left*: A Morse key as used by telegraph services *c.* 1850–70. (*Science Museum/Science and Society Picture Library*) *Above, right*: Artist's impression of TIROS 1, the world's first meteorological satellite, launched 1960. (*Science Museum/Science and Society Picture Library*)

The sinking of the *Royal Charter* off the coast of Anglesey, North Wales, 26 October 1859. (*The Illustrated London News Picture Library*)

James Glaisher with aeronaut Henry Coxwell in the basket of the balloon *Mars*, which took them 7 miles above the earth on 5 September 1862. (*The Illustrated London News Picture Library*)

Top, left: Francis Galton. (*Royal Geographical Society, London*) *Top, right:* Admiral Robert FitzRoy, RN. (*Image © K. E. Woodley*) *Above, left:* Matthew Fontaine Maury. Portrait by Matthew B. Brady, *c.* 1860. (*National Portrait Gallery, Smithsonian Institution*) *Above, right:* Joseph Henry. Portrait by Henry Ulke, *c.* 1875. (*National Portrait Gallery, Smithsonian Institution. Transfer from the National Museum of American Art*)

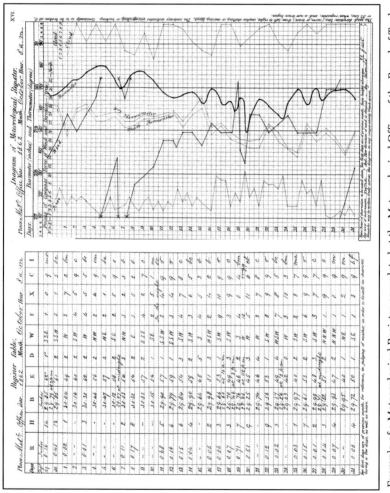

Example of a Meteorological Register completed at the Meteorological Office of the Board of Trade in October 1862, from which FitzRoy constructed his weather forecasts. (
National Meteorological Library and Archive)

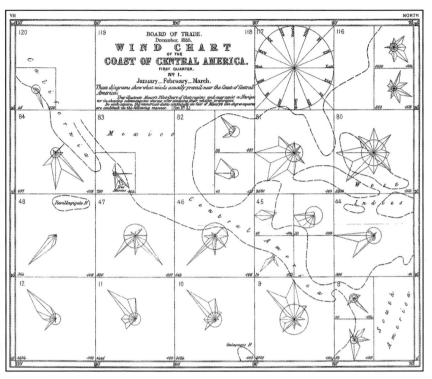

FitzRoy's Wind Chart of the Coast of Central America for January, February and March, drawn in December 1855. 'They illustrate Maury's pilot chart of that region, and may assist a Navigator in shaping advantageous courses after studying their *relative* proportions.' (*National Meteorological Library and Archive*)

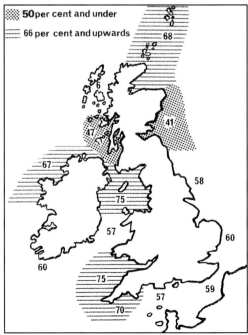

> ⣿ **50** per cent and under
> ═══ **66** per cent and upwards

Map showing the accuracy of FitzRoy's storm cautions, 1864. (*Bernard Ashley, Weather Men (Allman & Son, 1970)*)

METEOROLOGICAL REPORTS.

Wednesday, July 31, 8 to 9 a.m.	B.	E.	M.	D.	F.	C.	I.	S.
Nairn..	29·54	57	56	W.S.W.	6	9	o.	3
Aberdeen	29·60	59	54	S.S.W.	5	1	b.	3
Leith	29·70	61	55	W.	3	5	c.	2
Berwick	29·69	59	55	W.S.W.	4	4	o.	2
Ardrossan . ..	29·73	57	55	W.	5	4	c.	5
Portrush	29·72	57	54	S.W.	2	2	b.	2
Shields	29·80	59	54	W.S.W.	4	5	o.	3
Galway	29·83	65	62	W.	5	4	c.	4
Scarborough ..	29·86	59	56	W.	3	6	c.	2
Liverpool.. ..	29·91	61	56	S.W.	2	8	c.	2
Valentia	29·87	62	60	S.W.	2	5	o.	3
Queenstown ..	29·88	61	59	W.	3	5	c.	2
Yarmouth.. ..	30·05	61	59	W.	5	2	c.	3
London	30·02	62	56	S.W.	3	2	b.	—
Dover..	30·01	70	64	S.W.	3	7	o.	2
Portsmouth ..	30·01	61	59	W.	3	6	o.	2
Portland	30·03	63	59	S.W.	3	2	c.	3
Plymouth.. ..	30·00	62	59	W.	5	1	b.	4
Penzance	30·04	61	60	S.W.	2	6	c.	3
Copenhagen ..	29·94	64	—	W.S.W.	2	6	c.	3
Helder	29·99	63	—	W.S.W.	6	5	o.	3
Brest	30·09	60	—	S.W.	2	6	c.	5
Bayonne	30·13	68	—	—	—	9	m.	5
Lisbon	30·18	70	—	N.N.W.	4	3	b.	2

General weather probable during next two days in the—
North—Moderate westerly wind ; fine.
West—Moderate south-westerly ; fine.
South—Fresh westerly ; fine.

The first weather forecast supplied by Admiral Robert FitzRoy of the Meteorological Office and printed in *The Times*, London, 1 August 1861. (*News International Syndication*)

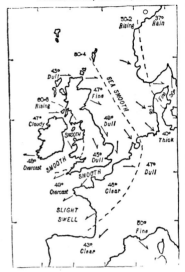

WEATHER CHART, MARCH 31, 1875.

The dotted lines indicate the gradations of barometric pressure. The variations of the temperature are marked by figures, the state of the sea and sky by descriptive words, and the direction of the wind by arrows—barbed and feathered according to its force. ☉ denotes calm.

The first weather map to be printed in *The Times*, 1 April 1875. (*News International Syndication*)

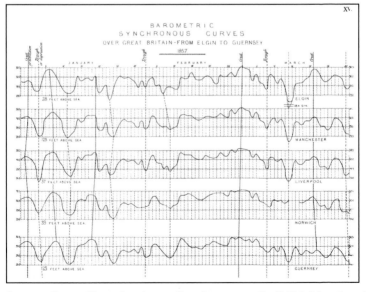

A Synchronous Curves Chart, showing readings from across the British Isles at a set time on a set day. Part of a series constructed at the Meteorological Office of the Board of Trade in 1857. (*National Meteorological Library and Archive*)

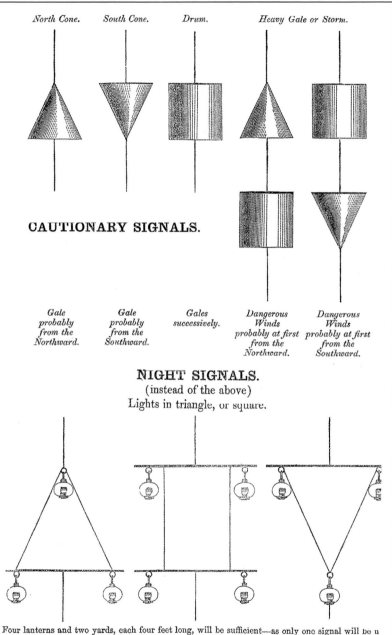

North Cone.　　*South Cone.*　　*Drum.*　　*Heavy Gale or Storm.*

CAUTIONARY SIGNALS.

Gale
probably
from the
Northward.

Gale
probably
from the
Southward.

Gales
successively.

Dangerous
Winds
probably at first
from the
Northward.

Dangerous
Winds
probably at first
from the
Southward.

NIGHT SIGNALS.
(instead of the above)
Lights in triangle, or square.

Four lanterns and two yards, each four feet long, will be sufficient—as only one signal will be u
at night.
These signals may be made with any lanterns, showing either white, or any colour, but *alike*.
Red is most eligible. Lamps are preferable to candles. The halyards should be good rope,

Cautionary storm warning signals devised by Robert FitzRoy, 1860.
(*National Meteorological Library and Archive*)

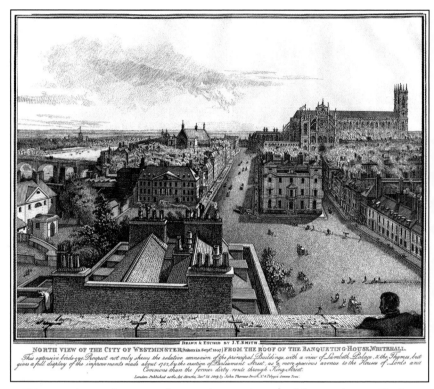

This engraving, made in 1807, shows Parliament Street (middle centre). No. 2 (the first home of the Met Office from 1854) was next to the nearest end of the street, on the right. The entire block between Parliament Street and the old King Street was later demolished in the 1860s to make way for the Government Offices that now line the north side of Parliament Street. King Street has disappeared beneath them.
(*City of Westminster Archives*)

PARLIAMENT STREET

A portion of Tallis's street view of Parliament Street, showing No. 2. Tallis made a series of street frontage views of various London streets in 1838–40.
(*City of Westminster Archives*)

goes into the making of a good physician. However, he soon discovered that Cambridge provided only theory and, to fulfil his creative bent, Galton turned to the invention of mechanical devices, writing his findings in verse like grandfather Erasmus.

He completed his studies in 1845, with little to show for them. Throughout twenty years of intense education, from infant prodigy taught by his elder sister, through stuffy, classical grammar-ridden schools to the examination-driven academia of the Cambridge system, nothing he had been forced to learn had matched Francis Galton's thirst for the practical application of science. Neither had it satisfied his wanderlust. Finally free from formal education, he wiped his hands of the medical profession, claiming that nothing he had been taught had provided a proper preparation for practice as a physician, and took himself off around the world. The year 1850 found him on a sailing ship bound for the Cape of Good Hope, from where he penetrated into a little-known tract of Tropical Africa, learning how to fend for himself as he went and mapping everything he came across. He took latitude and longitude observations, the latter by the lunar observation method, since he did not have a chronometer, although the observations were not very accurate, since his only instruction in how to make them came from the brief lessons given by the captain of the ship on his trip out.

The book he wrote of his travels, when he returned in 1852, earned him the Royal Geographical Society's Gold Medal and was snapped up eagerly by a book-buying public whose taste for armchair adventure had been fuelled by the explorations of Galton's contemporary, David Livingstone. Charles Darwin, now an invalid living in Kent with his wife and seven children, wrote in admiration to his cousin: 'what labours and dangers you have gone through . . . the objects of my study are very small fry, and to a man accustomed to rhinoceroses and lions, would appear infinitely insignificant'.

With the onset of the Crimean War, Galton's travels were curtailed. He took an avid interest in the campaign, though, devouring daily newspaper reports of far-off events. Like Robert FitzRoy, who chafed at the Navy's lack of foresight in the face of a storm, Galton cringed at the ineptitude of an army that sent soldiers into inhospitable territory totally unprepared for camping in the trenches. Generals, sitting in the comfort of a London club, believed, in typical Victorian fashion, that any foreign country can be treated and tamed as another England, so they sent their men forth to travel and camp in an alien wilderness, equipped only for a midsummer stroll across Salisbury Plain. Years of exploration had taught Galton the art of survival and he rushed to offer the Army a series of training lectures. To its credit, the Army did take Galton up on his offer, and he presented to selected officers a structured course on ways of finding food and water, creating fire, and making shelters, and advising them on how to cross rivers and march over trackless land. It all would have been invaluable, and would possibly have saved even more lives than Florence Nightingale and her band of nurses, but unfortunately the Army did not get around to holding the courses until the Crimean War was nearing its end. Whereas the Black Sea storm pitched FitzRoy into a career in meteorology, Galton's war effort came too late to make an impact and his military association was short-lived and soon forgotten.

At least his travel lust was sated for the time being, and he could turn his attention to science. His brush with the Army and its logistical problems had served to demonstrate the need for an easy means of distance communication. Some years before, he had attempted to design a printing telegraph he called a 'teletype', though the instrument itself never actually made it into manufacture. However, the memory of Scafell and the Ordnance Surveyors with their heliostat had never left him, and ever since he had toyed with the idea of a handheld version of the heliostat that travellers, soldiers and sailors could easily transport. At last, the design was perfected. It was published in the British Association for the Advancement of Science report in 1858, followed by a diagram in the *Engineer* magazine in October that year.

Francis was convinced, though, that the ideal application for the instrument was naval, an opinion supported by his cousin Charles. And Darwin it most probably was who suggested that his old friend, Vice-Admiral Robert FitzRoy, now head of the Meteorological Office but still with a good deal of influence at the Admiralty, might be the man to approach. True, they had not met for a year or so, and Robert had left their last meeting not in the best of tempers after a confrontation over Darwin's views on evolution. But then Charles well knew Robert for the grumpy old so-and-so he could be; and he also knew him to be forgiving, fair and generously disposed to lend a young fellow a helping hand.

So Francis Galton dispatched the fruits of his years of labour to Robert FitzRoy, trusting he would put in a word for him in the right quarters.

FitzRoy would have found Galton's invention easier to deal with had the concept been a folly, like many he saw, or had there been a major flaw in the design. But the Hand Heliostat was well conceived and accurately drawn and its usefulness hugely evident. What was more, it was a practical device. Galton claimed already to have made one for himself that was efficient up to 10 miles. The instrument looked rather like a telescope, adapted to catch the sun in order to make a signal. Through the viewfinder, a portion of the sun's rays was intercepted by a mirror and appeared to the eye as a small spot on the landscape. This spot marked the position in the distance where the signal would flash. It was simple and ingenious, and FitzRoy must have been impressed. Ordinarily he would have gone overboard with praise and support for such a promising invention, but this was far from being an ordinary case.

Francis Galton, famous adventurer and explorer, was now proving himself to be an imaginative inventor and skilled engineer, just like his illustrious grandfather Erasmus Darwin. Charles Darwin's tales of Erasmus's various exploits, foibles and amazing creativity would have entertained FitzRoy through many a dark hour in a cramped cabin at the lonely tip of South America. Seeing the old man's genius re-emerge in one grandson was proving enough of a challenge; to discern it again, perhaps more so, in this second would have made FitzRoy feel distinctly uneasy. An inveterate reader, with an avid interest in the different countries and cultures of the world, he must have read Galton's books. He would know him as the man who had resided with Muslims; the man who, far from viewing them as pitifully in need of spiritual guidance, had recognised that there could be more than one valid religion and that their beliefs were equally as effective as Christianity in helping them cope with the problems of life. He was also a man who wholeheartedly embraced the theories of his cousin and who regarded the teachings of the Bible as a nightmare of superstition. With the Darwin pedigree and with such anti-Christian tendencies, Galton was little less than the Devil incarnate in FitzRoy's eyes.

Now FitzRoy himself felt under personal attack from the man. What was Galton trying to prove, by sending this invention to him? It was not directly a meteorological instrument; its application was more obviously naval. Why not send it directly to the Navy? He forwarded the papers to the Admiralty, with the minimum of a covering note, and drafted a curt reply to Francis Galton merely stating that his application had been passed on to the proper quarters.

Fortunately, he had far more positive matters to occupy him, one of which was certainly stretching him intellectually. It was the newly available English translation of Henry Dové's book on the theory of storms, *Das Gesetz der Stürme*. The book was the universally accepted authority on the nature of storms, and FitzRoy simply had to understand Dové's thinking to be included in meteorological discourse. The translation had been carried out by a young Irish student, Robert Scott, studying meteorology in Prussia, under Dové himself. Therefore, the translation ought to have been accurate, yet FitzRoy, although he would not admit it, was struggling to make sense of it.

However, it was a different topic altogether that was providing the most stimulation, and it concerned two very distant points on the globe: Bermuda in the middle of the north-west Atlantic, and Halifax, Nova Scotia. These two had little in common, apart from sharing the same longitude (around 65°W) and both being the destination for a wind-speed measuring instrument called an anemometer.

Wind speed was normally denoted in the terms of the Beaufort scale. It was easy to use, needed no instruments and its accuracy was certainly sufficient for weather logs and the tables compiled from them in the Meteorological Office. However, scientists required something measurable by instrument, something they could give a figure to, and use in calculations, such as miles per hour. There were in existence instruments called anemometers that purported to be able to give a measurement to wind speed. The first anemometer had been invented at least four centuries earlier by an Italian, Leon Battista Alberti, and was a rotating perpendicular disc that inclined to the wind. Two hundred years later, Robert Hooke – often credited as the inventor of the anemometer – had used and improved the same design. Since then, the instrument had continued to develop, and in 1838, when he was addressing the British Association on his progress to developing a law of storms, Colonel William Reid had remarked:

It is very desirable that these beautiful instruments should be placed beyond the limits of our own island, particularly in the West Indies and at the Cape of Good Hope, where they may measure the force of a gale such as no canvass [*sic*] can withstand; that which forces a ship to bare poles . . .

It is not only to measure the wind's greatest force that it is desirable these anemometers should be multiplied and placed in different localities, but that we may try, through their means, to learn something more of the gusts and squalls which always occur during storms.

Colonel Reid's appeal did not bear fruit, but eighteen years later, in June 1856, the meteorologists of the Royal Society decided they would like to instigate a scheme, on the lines of Colonel Reid's suggestion, to aid the study of the movement of the atmosphere. By this time, a big breakthrough in anemometry had been achieved by Irish astronomer Thomas Romney Robinson, who in 1846 invented the cup anemometer, similar to the instrument we are familiar with today. Four hemispherical cups rotated horizontally to the wind, and, by a series of gears and wheels, the number of revolutions per minute could be measured. The addition of a cylinder covered with paper and a pencil on an arm meant that the apparatus could automatically record a continuous measurement of wind speed. Robinson claimed the cups moved at one-third wind speed, which was the accepted theory until later disproved in 1872.

The trouble with these instruments was that they were big, bulky, expensive and impossible to use on board ship. So a proposal was put together to set up self-recording anemometers at various locations around the Atlantic – Ascension Island, the Azores, Bermuda, Madeira and St Helena – from which it would be possible to obtain a series of regular and continuous observations. In October 1857 Lord Wrottesley had written to Lord Stanley, President of the Board of Trade, informing him that the instruments should be provided. Not surprisingly, Lord Stanley did not rush to implement this costly proposal. When Stanley retired, early in 1858, Wrottesley wrote again, this time directly to Robert FitzRoy, asking him to investigate progress, if any.

This was news to FitzRoy, for until then he was not aware of the Royal Society's plans for a network of anemometers. FitzRoy had himself long harboured the desire for a practical, handheld anemometer for use on ships, and now, with his increasingly scientific need for wind-speed measurements, the appeal of an instrument above a subjective eye-based observation was enormous. He had been watching with interest the modifications being made by W. Snow-Harris to an instrument invented in 1775 by Edinburgh physician James Laird, which used pressure on fluid in a periscope-shaped tube to measure speed. However, development was slow and results not proven to be accurate. When Wrotteley's appeal arrived, he could not believe his luck. If he could persuade the Board of Trade to part with the money, all these wonderful instruments planned by the Royal Society would come under his control. Quite legitimately they would provide him with extra sources of simultaneous observations, while acting at the same time as a front for his own private investigations into atmospheric waves.

He investigated and discovered that only one instrument had been ordered, and that it was destined for Bermuda. Lord Wrottesley found him most cooperative in helping to obtain sanction for more anemometers, but even their joint forces squeezed only one more instrument out of the coffers of the Board of Trade, for service in Halifax, Nova Scotia. In February 1859, on a ship loaded with two giant anemometers, FitzRoy's clerk, Babington, began a cruise to Bermuda.

Of his staff, Babington was the man with the highest technical aptitude, and a growing enthusiasm for the new scientific approach being taken by the department. In the rules of strict Civil Service hierarchy, Babington's grade was 'Junior Clerk', which, despite its rather subordinate sounding name, did actually make him the highest graded of FitzRoy's assistants. Ever since his first deputy had almost instantly been recalled, there had been friction between FitzRoy and the succession of appointed replacements. At the heart of the Meteorological Department were FitzRoy and his faithful trio of Babington, Pattrickson and Townsend. The draughtsman, Pattrickson, was the senior in years, so FitzRoy declared him to be deputy, an arrangement that worked happily until the Civil Service machinery ground round a cog, and gave Babington promotion over the head of Pattrickson. FitzRoy was incensed. Not that he thought Babington undeserving – it was obvious he rated the man highly – but he felt that strictly on merit no one was more deserving than Pattrickson, and he insisted that

Pattrickson remained his deputy and that all the other Meteorological Office clerks were junior to him, no matter what their Civil Service substantive grade might be. This did not worry the easygoing Babington, but it did upset the stream of died-in-the-wool Civil Servants who came into the department, and went again, refusing to work under the arrangement. As aggravations go, it was not a major one, but to FitzRoy, the naval officer, this civilian quirk of resenting his authority and reacting against it, no matter how fair he tried to be, was one of those niggling irritations he would rather have done without. Much higher in the nuisance scale, though, came the put-downs delivered to him by those in the outside world.

Not long after Babington had departed, FitzRoy received a piece of news that left him completely stunned. Captain Matthew Maury had been nominated to the French Legion of Honour for his work in meteorology. Excitedly, Maury wrote to FitzRoy implying that the honour fell not just to himself but to FitzRoy and all other meteorologists. 'You will I am sure be gratified to see . . . this additional evidence of the favor in which the good cause in which we are laboring is held by others.'

He was wrong – FitzRoy was not 'gratified'; he was jealous. Maury had had the good fortune to have organised the first international conference on the subject, yet he was no more than FitzRoy himself: a naval man with a passion for the weather. Both were marginalised by those who considered themselves to be true meteorological scientists. FitzRoy could judge from the free flow of ideas and information gushing from Maury in his letters that, meteorologically, he was not an original thinker. FitzRoy, on the other hand, with his rapidly developing theories on the atmosphere, felt he hovered on the brink of true scientific breakthrough. Since Maury considered the two of them were labouring together, FitzRoy felt it was time to exploit this relationship to meet his own ends.

Babington's imminent arrival off the coast of North America with his anemometers gave a good excuse for choosing now to broach to Maury the subject of simultaneous observations, making it seem as if he had only recently thought of it. With each anemometer, he explained to Maury, a weather station would be set up, to take observations at set intervals. From these observations and in conjunction with other readings obtained simultaneously at sea, the aim was to construct a series of synoptic charts covering the Atlantic. It would be even more beneficial, he suggested, to establish a multitude of observation stations on the east coast of North America, and he wondered, would Maury be able to help in finding volunteers?

He knew mention of the anemometers was bound to excite the American, since winds were his speciality. And they did, for Maury wrote back to FitzRoy at length on the subject of ship's anemometers, which was not quite the response FitzRoy had hoped for. As far as FitzRoy was concerned, the anemometers were by the by. His pressing need was for Maury's cooperation in providing assistance in the matter of simultaneous observations. Maury was far less effusive on this subject. Disappointingly – and probably because, for a quiet life, he did not want to bring the wrath of the Smithsonian's chief tumbling down upon himself once again – he told FitzRoy that, if he wanted to obtain additional readings from land-based American observers, he must make approaches through the proper channels.

Nothing daunted, FitzRoy used his own government position to make application to Washington for help with his observations. Permission was duly granted, and, with the request directed through official channels, Maury was then free to give FitzRoy all the help he wanted, and indeed he did. In June he placed advertisements for volunteer observers to help the English admiral in his task. Replies arrived by every post, from lighthouses, from harbour masters, from enthusiastic amateurs all along the coast – and, astonishingly, one from Alexander Bache himself. Thirty-eight suitable locations were selected, and FitzRoy provided packs of instructions, which were supplied to the volunteers.

In parallel with all this transatlantic activity, FitzRoy chose now to roll out into the open his plan for simultaneous observations. Implying that the construction of synoptic charts was a by-product of the anemometer exercise, he drafted a memo intended for widespread distribution. 'In connexion [*sic*] with the setting up of Anemometers, a series of wind charts is in progress which will show the state of the atmosphere over the ocean and its boundaries once a day (at least) during certain selected periods in the twelve or fourteen months of special observation . . .', he said, and went on to announce his study of the sequence of the prevailing weather arriving from across the Atlantic, over a period of one year, from September 1859 to September 1860.

His ambitious plan needed cooperation not just in America, but in Europe, too. Unlike his unsuccessful applications for wide-ranging observations just two years earlier, this one was hugely successful. He quoted the Royal Society's original 1856 request to the government 'to institute a series of regular and continuous observations in the Atlantic Ocean', and it was amazing what a little bit of backing from the Royal Society could achieve – or perhaps it was due to Thomas Milner Gibson's lead as the new President of the Board of Trade. Whatever, he was given immediate approval to send his memo to His Britannic Majesty's Consul in Norway, which at first sight may not appear to be a large territory, but, considering the Consul's representation extended over Iceland, Denmark, Hanover, Heligoland, Holland and the Farne Islands, there was a large range of potential areas for observation. FitzRoy followed up this success with a request to distribute his memo to various coastal authorities in England, Scotland and Ireland, plus Lloyd's agents and the Cunard manager. That, too, was given immediate approval.

It did not appear to bother him that, when all these observations came rolling into his office, the exercise of converting them into synoptic charts would swallow up copious hours of his clerks' sparse time. As far as he was concerned, it was appropriate anyway to scale down the tabulation of data that was his department's *raison d'être* – 'the great collection of observations that is now being made . . . should not be allowed to extend indefinitely lest the data should grow overwhelming', he said, baldly stating an opinion he had already intimated in his 1858 Report. The reason it was overwhelming was perhaps because he did not apportion sufficient time, or consider streamlining his procedures for dealing with it, but this fact appears to have been conveniently ignored. That it was not being ignored in other circles and by certain other scientists was something that FitzRoy would learn to his cost. But that lay in wait for the future.

For now, the way ahead shone clear. FitzRoy had finally managed to decipher Henry Dové's theories on the behaviour and causes of storms. Enlightenment crashed into his atmospheric-wave theory with cyclonic force, clearing away his doubts, brushing out the inconsistencies and leading the way to a brand new understanding – plus a sudden interest in matters aeronautical.

Atmospheric Gyration

One Monday in August 1859, at 1.30 p.m. precisely, Robert
FitzRoy joined a group of dignitaries in the yard of the
Wolverhampton Gas Company. A large wicker basket stood
incongruously on the coal-blackened cobbles, surrounded by
billowing yards of silk. Supervising proceedings was a Mr
Green, balloon aeronaut, whom FitzRoy would have known by
sight. He had been present at the meeting of the British
Association's Balloon Committee at Burlington House in
Piccadilly in July, which FitzRoy had himself attended. FitzRoy
had recently joined the committee along with James Glaisher,
another new member attracted by the possibilities offered to
the meteorologist by a gas balloon.

Glaisher might have been new to the committee but, unlike
FitzRoy, he was not new to ballooning. As meteorologist
attached to the Royal Observatory at Greenwich, he must often
have envied the astronomers gazing starwards through their
telescopes and wished that it was equally as easy to point a
glass at the sky to observe the upper atmosphere. Stuck to the
surface of the earth, a meteorologist had the odds stacked
against him in trying to make sense of a three-dimensional
entity with only two dimensions to play with. That the upper
air had a life of its own was frustratingly obvious, especially
on a breezy day, when scattered cumulus scudded briskly
across the sky while, above them, high cirrus glided in the
opposite direction, seemingly oblivious of the wind's pull.
Glaisher had gamely climbed mountain summits in the name
of research, but in terms of the earth's atmosphere these were
but pimples. So in 1852, when the British Association had
sponsored Mr Welsh of the Kew Observatory in a series of
balloon ascents to take meteorological readings in the upper
air, Glaisher was there to watch, the keenest of spectators.
From his position on the roof of the Greenwich Observatory,

he could follow the entire flight with his telescope, from take-off at Vauxhall in London, to touch down some 70 miles distant near Folkestone on the Kent coast. Glaisher's ambition was to zoom up into a cloud himself, to feel it close around him, to look down on it from aloft and to observe, measure and experience everything he could. Now that the British Association was planning further excursions into the air, he had joined the committee with the aim of riding with the aeronauts.

However, it was not Glaisher who was to go aloft that afternoon from the Gas Works yard in Wolverhampton. The British Association had decreed that the privilege should go to a young man and competition for this had been intense – two young medical students from Glasgow University and a professor, Professor Tyndall, himself a member of the Balloon Committee, had all vied for the position, but Mr Storks Eaton, who had been looking after the Kew Committee's instruments in storage since the last ascent, had offered his services free and had received the honour.

Glaisher, though, had made the trip to Wolverhampton that August day, as had Colonel Sykes, the Committee Chairman, and Lord Wrottesley, FitzRoy's erstwhile champion. Wrottesley, himself a member of the Balloon Committee, had considerable estates just outside Wolverhampton, and he had persuaded the owners of the local Gas Company to supply the gas for the balloon. When the balloon was laid out and ready, the gas began to flow, but progress was painfully slow – filling the balloon with coal gas was a dangerous and painstaking task that could not be hurried. By late afternoon, what had begun as perfect weather deteriorated when the wind started to gust in the yard. After 63,000 cubic feet of gas had been delivered, the balloon still lay across the yard like a prostrate whale, and with evening fast approaching, it was decided to postpone the ascent to the next day.

The visiting members of the Balloon Committee retired to the hospitality of Wrottesley Hall, and when they returned at 1.20 the following afternoon, the balloon was already upright, swaying majestically in front of the Gas Works buildings. But just as the party was passing through the gateway, a sudden gust of wind snatched at the neck of the balloon. The material flapped violently, the silk tore and gas gushed from a split several yards long. Mr Green declared the balloon to be damaged beyond immediate repair, and the attempt was abandoned.

While Glaisher's hopes were no doubt raised at the prospect of flying in the next attempt, FitzRoy's designs were totally earthbound. He would be quite content with second-hand information, though he did desperately need the evidence of reliable observations in order to cement the theory of the atmospheric model that was now exciting him. No date had been set for a further flight, since it had to be sanctioned by the full British Association. FitzRoy could not wait. He was impatient to set his theory, if not in blocks of stone, then at least into the permanence of the printed word. He went ahead and published an ambitious little paper, which he entitled *Notes on Meteorology*.

The *Notes* offered a total repudiation of the atmospheric-wave hypothesis. Dové's reasoning had quashed FitzRoy's own belief in this theory – yet some people, he says, disassociating himself from them, regarded the atmosphere as behaving like the sea, with crests and troughs. Balloonists, he countered, had proved that there are different layers and currents in the atmosphere. If you have waves of air sloshing about the earth's immediate surface, without causing similar waves in the upper air – which it appeared did not happen – then wave troughs in the bottom layer would make gaps between layers, which was impossible. Anyway, he pointed out, the 'trough' represents an area of low pressure, which it had already been proved is actually an area of rising air, light and at its most expanded.

He presented, instead, a theory of atmospheric circulation, which he credited 'on the authority of Herschel and Dové' with a little bit of FitzRoy thrown in. It was known in FitzRoy's day that air pressure over the equatorial regions is permanently low and over the tropics and poles is permanently high, while it is in the temperate zones that pressure, and the weather, change unpredictably. A theory of atmospheric circulation has to take account of this phenomenon, and Dové's Law of Gyration, it seemed to FitzRoy, fitted the bill to perfection. Air, he explained, is a mixture of gases, and, in the same way that coarse shot can absorb small shot and sand without growing in bulk, air can absorb water into the interstices between its particles without increasing in volume. Heat, however, was another matter. No scientist had yet managed to define it – the best FitzRoy could manage was 'that mysterious and subtle agent which acts mechanically and in an unexplained manner'. But its effect on air was well known. In some way, it was able to insinuate itself between the particles of air, expanding the space between so that there was more room for moisture to be absorbed, and making it lighter so that it would rise.

FitzRoy's explanation of the theory of atmospheric circulation ran basically on these lines:

Over the equator, the air is always warm and therefore constantly rising, keeping the mercury low. As it rises, denser, cooler air must fill the vacated place at the surface, and this cooler air is drawn steadily from the north and south poles to replenish the supply – the polar wind. In its journey to the equator, the polar wind is warmed and picks up moisture from the oceans until it, in turn, rises on the equator. However, the equatorial upflow cannot rise infinitely, or the earth's atmosphere would leak out into space. At a certain height – and the guess was 10–12 miles, though it had to be a guess as no experimenter had explored above 5–6 miles – the air is deflected back towards the pole, moving across and above the polar air flowing in over the surface. As the equatorial air rises and travels, it becomes cooler and heavier and deposits its moisture over the tropics as rain. At around 30° north and south, the 'horse latitudes', this air has become dry, and some of it descends to the surface, causing the clear skies and high pressure of the desert areas. From here, it returns to the equator while the rest remains as upper air, continuing its journey to the pole. In the cold of the pole, the mysterious 'heat' element escapes from between the air particles and leaves for some unknown destination. The air is now dense, and falls to earth, making the mercury stay high. Inexorably, the pull of the incipient vacuum at the equator draws the air, and it begins its journey again, back to the warm. Temperate latitudes are ruled by alternating currents of air, sometimes polar and therefore cold, sometimes equatorial and therefore warm. When the two meet, they whirl together into cyclones, and where the clash is severe, they toss and buffet against one another in the turbulences we call storms.

The basic model is disturbed by other factors. Some are localised, such as mountains, lakes and the sea's currents.

But the biggest force acting upon the model's north–south flow is the earth's rotation and its spherical shape. The earth makes a complete rotation in one day, no matter where on the surface you happen to be. If you are standing at a spot on the equator, you will be whisked round briskly at a speed of over 1,000mph, in order to cover the near 25,000 miles of the circumference in a day. However, the nearer to the poles you stand, the smaller the circle your day's transit will describe, so the speed at which you rotate through space is much slower. Now imagine you could leap from the surface at the equator and hover, independent of the atmosphere. With no air current to accelerate you and no friction to slow you down, you would maintain the same speed as the spot you left, allowing the ground below to appear to be stationary beneath you. Now, propel yourself northwards. As you look down, you will see you have ceased to hover. Since the ground is now moving at a slower pace than yourself, you will seem to be speeding in an easterly direction. When you land, you will be not just north, but both north and east of where you set off, and it will appear to the people you land among that you have suddenly sped in from the south-west.

This is exactly what happens to the air. As the warm air leaves the equator, it continues to rotate at more or less the same speed as it did near the ground. However, as it travels north in the northern hemisphere, the ground beneath it begins to move more slowly in relation to it, so in earthly terms, it is approaching from a south-westerly direction. Conversely, returning polar air travels from north to south maintaining the slow rotational speed of the poles, while the ground under it moves faster and faster as it moves south, causing the polar wind to blow across it from the north-east. In the area between the equator and the tropics, this 'polar' airstream is the lower atmosphere, and in these regions the wind at the surface is reliably constant. Centuries of sailing vessels plying their trade around the world knew and took advantage of these winds, giving rise to the name Trade Winds, winds that to the north of the equator blow as north-easterlies, and to the south as south-easterlies.

It is a complex proposition to understand, so little wonder that FitzRoy had trouble getting to grips with it. For his trouble, FitzRoy was to feel the lash from Sir John Herschel, who had expounded on the subject of atmospheric waves in his definitive entry on meteorology in the recently published *Encyclopaedia Britannica*, and therefore towards whom FitzRoy's criticism had been most obviously aimed. FitzRoy had met John Herschel at the Cape of Good Hope when the *Beagle* had called on her way home from the South Seas. Correspondence between the two had continued in fits and starts over the intervening years. FitzRoy seems to have regarded Sir John with a certain degree of awe, honoured that the esteemed scientist should allow himself to be used as the sounding board for many of FitzRoy's emergent meteorological theories. Sir John was not the most sympathetic nor encouraging of mentors, often castigating FitzRoy for his efforts, but FitzRoy, time and again crushed by these between-the-lines put-downs, shrugged them off and rose once more, flattered to be given the ear of the great man. Now, for once, it was Sir John who found himself at the receiving end, his own theories castigated by the man he considered his scientific inferior. In retaliation, Sir John sieved through the logic of FitzRoy's theory. Despite his best efforts, though, he could find little to pick on other than FitzRoy's terminology. Why did he keep using the seaman's phrases of 'easterly' and 'westerly'? A landsman, said Sir John, does not know whether 'easterly' is coming from the east or going to the east, so how will he manage to grasp what FitzRoy is trying to explain? If that was the worst he could come up with, FitzRoy would not have lost any sleep in fretting about it.

In typical FitzRoy fashion, though, he did not confine himself to theories of atmospheric circulation. He wrote of other influences on Britain's weather, arguing that its climate very much depends on the Gulf Stream and the warm, moist westerlies it brings with it. Somewhat perspicaciously he added that, should this 'peculiar circumstance' alter, there is 'fear of a gradual change in our average climate'. One reason why the Gulf Stream might possibly fail he envisaged as a 'diminution of ice in the polar regions', which a century and a half later is a very real fear indeed.

He pointed out what to him seemed a basic fact of human nature, that while it was all very well for his department to accumulate materials for scientific philosophers, unless he could give the people collecting them some very real reward for their labours, he could not expect their willing cooperation to continue indefinitely. 'Among the objects to which meteorologists should officially devote themselves seems to be the selection of practically and immediately useful points in the first instance, and purely scientific ones, for future objects, subsequently.' And what were these practical points, and for whom did he envisage they might be useful? Reid and Redfield, James and Burgoyne had just one aim in mind – to understand the nature of storms, so they could be prepared for or avoided. Maury's primary objective in measuring the winds was to point the sailor towards the quickest routes across the oceans. But FitzRoy's statement wreathed these worthy aims with modesty by comparison. If you watch the trend in temperature and pressure, together with the humidity, he declared, 'you will scarcely be mistaken in knowing what kind of weather you are likely to have for the next two or three days – which for the gardener, farmer, soldier, sailor and traveller must be frequently of considerable importance'. Here, for the first time, FitzRoy set up signposts to where he wanted his work to lead. He aimed to predict the weather, all weather, beyond simply winds and storms, and for the benefit not just of sailors but for the whole population. His aims were ambitious, more ambitious than he could ever realise, and he was determined to achieve them.

His statement was not original. It echoed almost word for word the writing of Lavoisier in France eighty years earlier. Lavoiser and Lamarck had not failed; they had merely succumbed to the French political climate. Now, a new disciple was preparing to follow their example. Lavoisier had stated that the art of weather prediction followed certain rules and principles, which required the full attention of an experienced practitioner for their interpretation. FitzRoy was that experienced practitioner. He had mastered the principles and now his vast, year-long, simultaneous-observation experiment would soon be providing him with the rules. One of these was already emerging.

Soon after a few of the earlier synoptic charts were partly filled, it became apparent that while there are *various* currents, sweeps, or circuits, in any given area of our temperate zone, intermixed and incessantly moving – the whole body of them (as a connected group), the entire mass of atmosphere in our latitude, has a constant, a perennial movement toward the east, *averaging* about five miles an hour.

Like many of FitzRoy's theories, it was not totally original. The proposition had first been put forward in 1856 by American William Ferrel, but it appears that few people in Europe had picked up on it. To FitzRoy, though, in conjunction with Dové's Laws, it was to form the basis of his understanding of atmospheric gyration.

He stood poised on the verge. It would take a calamity to divert him, the most determined of individuals, from his course. So what was to prove so momentous that it made him lose interest in his much-fought-for transatlantic simultaneous observations, just weeks after the trial had started, and sent him charging off in a totally different direction?

THIRTEEN

Royal Charter Storm

The stench from the Thames was ever vile, but, charged by an overheated atmosphere, fumes settled over Westminster in a viscid cloud. The long, hot summer of the previous year, 1858, had been named 'The Great Stink', and this current summer threatened to be stinkier still. The temperature passed the 90° mark, day upon day. Flies flourished, parasols fluttered and people wilted. Scented kerchiefs, pomanders, sprigs of lavender and posies of violets were the essential London fashion accessory, but really they made little difference.

The River Thames, which at the beginning of the nineteenth century had still supported freshwater fish, salmon included, just fifty years later was a sterile, noxious brew of raw sewage and rotting waste. It was the affluent upper and middle classes who were to blame. An increasing fastidiousness, along with Thomas Crapper's newly invented water closet, had overwhelmed the cesspits, so that they overflowed into the scanty rainwater drains under the streets and thence into the river. In the summer of 1858 high pressure had sat over the city for weeks on end and a hot sun had streamed down from a cloudless sky, cooking the excrement until an indescribable stench rose in through the open windows of the Houses of Parliament themselves. Politicians and civil servants hid behind curtains of sacking soaked in chloride of lime, and tons of carbolic were tipped into the river, but the stink refused to retreat. Such was its strength, it managed to achieve in a matter of days what years of lobbying had failed to do – a Bill to provide the funding for a new sewerage system for London.

The system that was subsequently built is still functioning today, keeping the Thames clean and the streets sweet. However, it was far too far into the future to provide relief for FitzRoy, returning to work after the aborted ballooning episode in Wolverhampton. Attempting to work in the heat and the reek must have been nigh on impossible and no doubt he was looking forward to getting away again – which fortunately he did, and very shortly, too, when he boarded a train at King's Cross station and let the GNR take him north on the long journey to Aberdeen.

The gentlemen of the British Association liked to leave London en masse in the summer – and with such a stink, who could blame them? – to hold their annual meetings at various locations around Britain. This year, it was Aberdeen's turn to host the twenty-ninth Annual Meeting. Robert FitzRoy was making the 1,000-mile round trip in order to mix in royal company, since the meeting was to be presided over by the association's President, Prince Albert, Queen Victoria's Prince Consort.

As FitzRoy settled in his seat to study the agenda, one particular item would have caught his attention. It was a motion, proposed by Colonel Sykes (he of the Balloon Committee), that the association resolve to make a representation to the Board of Trade to fund a system of telegraphed storm warnings. Since he had met the colonel only a month previously, it is highly likely that the subject would have been discussed by the two of them. After all, if the resolution were passed and the Board of Trade agreed, then it would be FitzRoy's Meteorological Department that would be responsible for putting the scheme into practice. Therefore, he had a vested interest in assuring himself that the proposed scheme was workable and met his own criteria.

The motion was being proposed by Sykes on behalf of a certain John Locke of Dublin, writer, amateur weather researcher and contributor of many articles to the Royal Dublin Society. Locke had been in correspondence with Matthew Maury and from him he had learned of the system of telegraphed weather reports that had been running successfully in America for three years or more, and their intended use as storm warnings for the east coast. He proposed that a similar system of telegraphed storm warnings should be instituted in Britain.

It might seem strange that FitzRoy himself had not been pressing for such a system. He had, after all, hinted at the possibility nearly three years earlier when attempting to justify his simultaneous-observation experiment to a dubious Board of Trade. What could have been more helpful to the fishermen and sailors of small boats, whose interests he professed to have at heart? No more scratching their heads in front of a barometer, trying to guess the weather for themselves – they could rely on the information winging its way over the wires.

Storm warnings, such as they were, were at that time gleaned from the current weather conditions as observed *in situ*. In other words, the ship's captain at sea, or the harbour master on land, was the person best able to judge what the coming weather might be, from a combination of regular observations plus the look of the sky, the state of the winds and seas, plain common sense coupled with experience – and FitzRoy's *Barometer* book. If, by some miracle, the observer could be made aware of the weather way beyond his vision, the odds would be multiplied in favour of more accurate prediction, further in advance. In America and mainland Europe, that was the miracle the telegraph offered – although it had not yet been developed into storm warnings.

FitzRoy's apparent lack of interest stemmed not from lethargy or ignorance, but from what he deemed to be the system's impracticability in Britain. And that was all to do with geography. In eastern USA, the prevailing weather came across the land. Storms heading for the east coast would be experienced inland hours if not days before, and their progress tracked by many land-based observers. All that was needed – though granted this did require a certain amount of meteorological skill – was to be able to calculate which part of the coast they would hit, when they would arrive, and with what strength.

Under a similar system, the small coasters off the west-coast ports of Britain and Ireland would reap no benefit at all. Nearly all storms approached from the west, across the Atlantic Ocean, where the telegraph was powerless to help. Until there was a way of telegraphing from ships in the Atlantic (which would have to wait for another fifty years, until Marconi invented radio) there could be no advance warning of the actual fact of an approaching storm. And even if, when a storm did hit the west of Ireland, the information were telegraphed on across to the east coast of the mainland, because of the narrowness of Britain and the inadequacies of the telegraph service, most of the storms would have swamped the east before the message managed to arrive.

To set up a system in Britain on the proposed American model was decidedly not what FitzRoy wanted to do. To him, it was totally foolhardy. Warnings that time and time again arrived after the storm would be derided and ridiculed, and, though he might have longed for fame, it was not as the butt of a music-hall joke. However, even without radar and radio, storms did announce their coming. The warning was carried in the weather many hours ahead – air pressure, temperature, humidity, wind speed, rainfall and cloud cover all told their story. The only way to implement an efficient and accurate system of storm warnings in Britain was to be able to use barometer and other instrument readings in order to make a reliable estimation. To predict the weather, in other words. No one in an official position such as himself had yet presumed to issue weather predictions and, although he had already declared his hand in this – that he saw it as a practical proposition in the way that Lavoisier and Lamarck had – he was not yet ready to steam ahead with it. To feel more secure in his prognostications, he needed more knowledge of weather patterns. In other words, he was waiting upon the results of his year-long collection of synoptic data.

To FitzRoy's dismay, though, the members of the British Association were won over by Locke's arguments. They could not be blamed for this, since there was no detail of its means of operation in the proposal and, of course, it did appear to be an eminently worthy cause for a scientific body to back. The motion was accepted and the resolution passed. The British Association would send an application to the Board of Trade for funding for a system of telegraphic storm warnings. FitzRoy had much to ponder on his train journey back to London. Would the British Association's approach to the Board of Trade be merely a vague proposal, which was quite unlikely to attract funding? Or would it take the form of a specific system? And if it did, would the association simply go with Locke's design based on the American model, or would his own advice be sought? And if so, what would he say? Was he prepared to gamble his future on making predictions now, at this stage, when he was not entirely ready?

He did not know it, but his dilemma was to be resolved by two factors that had not even entered into his reasoning. Both were about to descend upon him with the shock of a double hammer blow. The first of these was the weather itself.

Unseasonably warm weather continued on well into the middle of October, until, on the night of the 19th, the temperature in London plummeted suddenly to well below freezing, while in Belfast the night remained as muggy and oppressive as ever. To FitzRoy, the east/west contrast spelled trouble, and he waited for it. However, the barometer height held, until the morning of the 22nd, when the expected drop occurred, deep and very worrying. FitzRoy predicted there would be gales and even snow in the north of England that day – though looking out onto a clear, dry London morning, it would have been hard to believe. That evening, he went to King's Cross to meet relatives due to arrive from Yorkshire, and spent a good hour or so kicking his heels. When eventually their train came in, they disembarked full of tales of the heavy snow and north-east gales that had delayed their journey.

Even so, the weather in the west stayed warm and humid. FitzRoy believed it threatened trouble, and in Holyhead, on the Isle of Anglesey off the north Wales coast, the people would have agreed with him that disaster was looming. Not that they saw it in barometer graphs or weather charts. For them a much surer sign was the meteor that, on the night of 25 October, shot across the heavens, and was seen way out to the west, for many miles out at sea. In the early hours of the 26th, a tremendous storm blew up. The next morning came the news that the steamer, the *Royal Charter*, had sunk. An iron ship only 4 years old, the *Royal Charter* had been constructed specially for the Liverpool–Melbourne route. Now, just a few miles short of Liverpool on her return from Melbourne, the ship, her captain and nearly all of her 500 passengers and crew had perished in a storm – a storm that FitzRoy himself had been able to predict.

The shock waves from the sinking resonated throughout the land. The papers splashed the news, and an entire nation went into mourning. It was the Black Sea disaster all over again, and everyone wanted to know why it was we still had learnt nothing. Where was all the fine talk about predicting a storm, avoiding disaster by watching the weather? It escaped the majority that a storm-warning system would not have helped the *Royal Charter*. She was heading into, not out of, port and, with no shore-to-ship radio, she could not have been saved that way. In actual fact, the knowledge of how to recognise and avoid that storm was already available to the captain of *Royal Charter*, yet he appeared to have ignored it. It was, though, used to good effect by William J. Johns, commander of the American sailing ship the *William Cummings*, which had been in close proximity to the *Royal Charter* on the night of the storm.

Johns's report, sent to Matthew Maury and later quoted by FitzRoy, stated: 'Having had many threatenings of bad weather for several days past, I began to apply your views as to storms . . . I did not suffer *any damage whatsoever* [FitzRoy's emphasis] more than usual in ordinary blows; only a little chafe and some spray.' The 'views on storms' alluded to were in instructions written by Maury which reflected the accepted wisdom at the time, stated by FitzRoy as: 'The simple rule of seamanship is when facing the wind the centre of the storm will be to the right . . . therefore you should go left.' The conclusion, as drawn by FitzRoy, is of 'a ship managed in accordance with instructions published for seamen, being saved, while the other, which adopted a different course, was lost'.

Perhaps as an ironclad steamer rather than an old-fashioned wooden sailing vessel, it was thought the *Royal Charter* did not need to heed such antiquated advice. The general public, however, were oblivious to the niceties of the arguments. Action was demanded and the British Association, sitting with a resolution to pursue storm warnings in its hands, was bound to present its proposal to the Board of Trade without delay, earning itself plaudits as a scientific body that could respond speedily to a crisis. And the Board of Trade, with a similar aim in mind, would grant the necessary funding, few questions asked. Then to FitzRoy would fall the task of making the system happen.

If he harboured any reservations, if he felt any reluctance to turn his full attention away from his grand scheme of simultaneous observations, they were about to be swept aside. On the night of 2 November, just one week after the devastation of the first, a second savage storm vent its full force over the battered British Isles, bringing yet further disaster to shipping. The case for a system of storm warnings was all but proved, and, if FitzRoy wanted to ensure it was based on predictions – his predictions – rather than a series of inadequate telegraphed reports, then he needed to act decisively in order to prove his ability.

People began to play into his hands. Observers, amateur and official, from across the length and breadth of the country, and from Europe, too, had been as intrigued by the unusual weather conditions as he had himself and, at his request, were soon inundating the Meteorological Office with instrument readings taken before, during and after both ferocious events. If he and his team worked quickly – and they already had both the means and a system worked out – he could have synoptic charts drawn, printed and circulated in a matter of days. The events were local and topical. The impact would be immediate, far greater than any number of dry, retrospective synoptic charts of commonplace weather events out in the north Atlantic. With these *Royal Charter* charts, he would be able to say, look, this is how a storm brews and this is how I can predict its coming.

He foresaw the making of his reputation, the imminent coming of the hour when he, Robert FitzRoy, would be hailed as the next public hero; the man who could pre-empt the vagaries of the weather by the application of science; the man who would save countless lives at sea.

Unfortunately, the second hammer blow was about to fall. Someone else was poised and ready to step out before him into the limelight as the next national sensation. Charles Darwin's *The Origin of Species* landed among the intellectual circles of middle Victorian England like a firecracker falling onto kindling. It left no room for waverers. There was no fence on which to sit. You either took on board the theory of evolution by natural selection as expounded by Darwin, or you remained a believer in the biblical account of the creation. You were either evolutionist or creationist, and now, quarter of a century on from the *Beagle*, Robert FitzRoy had transformed himself into the extreme reactionary creationist.

The ultimate, painful irony for FitzRoy was that the man responsible for letting loose the heresy had, in part, been himself. If he had not taken Darwin as his messmate, shared with him his cabin, his hospitality and his friendship over five long years at sea, then possibly none of this would have come about. But what happened had happened, and he could not lose the fateful bond that tied him to Darwin. *The Origin of Species* was not going to die. 'Evolution', 'Darwin' and 'the Beagle' were to be forever intertwined and, for the time being at least, the name 'FitzRoy' with them. It was a double insult to FitzRoy. Not only was he being linked inextricably with an idea and a concept that was utter anathema to him; his bid for recognition as a meteorological pioneer was sinking as fast as the *Royal Charter* itself, under the fickle tide of public allegiance now lapping on the shores of evolution. His attempt to gain acceptance among the ranks of the respectable and respected leaders of science had been hijacked. Few would want to talk of synoptic charts. Instead everyone would want to ask him about Darwin.

If he was to propel his boat out to the forefront of public opinion, then storms, and predicting their coming, were the way forward. In order to predict, it was time to harness a new force – the power of new technology.

Storm Prediction

The electric telegraph was the 'computer' of its day. It triggered a revolution in Victorian science and commerce in just the same way as, 100 years later, electronic logic circuitry would begin to transform the second Elizabethan age. At the beginning of the nineteenth century, when Robert FitzRoy was born, the speediest method of communication was achieved by four-legged horsepower, a method little changed since the time of the Romans. An urgent message from London to Exeter, a distance of 175 miles, it would be galloped in stages over the rutted turnpike roads, and seventeen hours later, potholes and highwaymen permitting, would arrive at its destination. There did exist a very limited system of semaphore telegraph – large poles with painted arms, standing on hill crests like a procession of wooden giants, by which messages were passed by line of sight, one peak to another – but that, like hilltop beacons, was a strategic signalling device, not available to the general populace. The coming of the railways in the 1840s and 1850s speeded travel, and with it the mail, but it would still take several hours to steam from London to Exeter.

The first electric telegraph in Britain was installed in 1839 beside 13 miles of Great Western Railway track between Paddington and West Drayton. By 1860, thanks to the network of telegraph cables spidering across the country, the age of instant communication had arrived. If FitzRoy, for instance, wanted to send a message to Exeter, he could take his message to the telegraph office, where a skilled operator, the human equivalent of a modem, would transform his words into Samuel Morse's code of dots and dashes, and tap, tip, tap them onto a telegraph wire, to be reassembled into words by another operator just a few minutes later, in Exeter. Likewise, he could receive messages from all over the country, and Europe, at the speed of an electric spark.

The world's first underwater cable was laid under the English Channel by the paddle steamer *Goliath* in August 1850, and, despite the fact that a French fishing boat fished up a length of the cable a few days later, communication between Paris and London was soon established. On 16 August 1858, thanks to Matthew Maury's identification of the 'Telegraph Plateau' beneath the surface of the Atlantic, the entrepreneurial aplomb of American businessman Cyrus Field and the skills of 26-year-old British engineer Charles Bright, Queen Victoria exchanged the first transatlantic telegraph message with American President James Buchanan. Unfortunately, the cable was damaged not long afterwards, and would not be fully operational for another few years, delayed by the intervention of the American Civil War.

Still, in the space of twenty years, a revolution had taken place and the telegraph, like the Internet now, had become a fact of everyday life. FitzRoy was not slow to realise the implications of this new technology.

On the other hand, he was circumspect in its uptake. The French might appear to have taken a giant step ahead of him with the transmission of weather reports, but FitzRoy's plans for simultaneous observations had not encompassed real-time data, and his clerks were making no headway against the constant flow of backdated registers as it was. Tabulation was years behind time. No point in flooding them with more. However, the *Royal Charter* storm, coupled with Darwin and his *Origin of Species*, were to mark the point where he allowed the new technology to take the Meteorological Office through into its next metamorphosis.

Where he had been lukewarm before, he was now fired with urgency. It was vital that the British Association's resolution should be taken up by the Board of Trade, for the principle of telegraph warnings and the infrastructure it required would suit his plans perfectly. It was also vital that he should succeed in getting implemented his own ambitious system of predictive warnings, rather than John Locke's more factual alternative, and then in proving that he could make it work. What he was proposing was revolutionary, a technique of storm prediction that no one before him had dared to attempt. Robert FitzRoy had never been slow to rise to a challenge, especially when the safety of sailors, seamen and fishermen was at stake. As he put it:

It is impossible for those who have studied practical meteorology, and know what power is now available for diffusing knowledge by telegraphy, not to feel a keen consciousness that many lives risked in gales or storms might be saved by simple modern appliances employed extensively.

He might have been risking his own reputation, but on the other hand, if he succeeded, he would stand in his own right in the full glow of public acclaim, rather than in the shadow of the mighty Charles Darwin.

On 5 December the British Association submitted its formal application to the President of the Board of Trade, and, as soon as he was aware of its arrival, FitzRoy hurriedly submitted his proposals to James Booth, Joint Permanent Secretary to the Board of Trade. A tense two weeks followed for FitzRoy while he awaited the outcome. He must have been impossible to live with, his usual sheen of charm and equanimity worn thin – snapping at his staff, prickly with his wife and ranting at every mention of the fashionable topic of evolution. At last, on 17 December, came a decision from the Board of Trade. Vice-President William Cowper asked FitzRoy to prepare a plan for conducting an experimental trial of telegraphed storm warnings. Victory! A cleft had opened and his way forward was clear.

FitzRoy spent that spring in careful preparation. He had already decided on the outline of the operation: from information relayed daily by telegraph to his offices in London, he would draw up synoptic charts, such as he had prepared after the *Royal Charter* storm; they would show him as surely as the *Charter* charts had if and when a storm was brewing; he would then send warnings directly to where they were needed well before the bad weather arrived. There was, though, much detail to fill in. For a start, who would take the observations, where and when? With the need to process data daily and within a turnround time of just a few hours, it was impossible to use all the observers who currently returned information to his office. Then there was the problem of carrying messages from the observers to the telegraph stations. In 1860 telegraph offices were not plentiful, and, being designed for commerce, they were not necessarily located exactly where FitzRoy would have wished. Nor were they, like those in France, government owned and run. It would not be as simple for him as it had been for Le Verrier to recruit the telegraph service and its operators to his cause. As a private enterprise, a telegraph company needed to make money – which meant each message would cost money. What sort of budget would he be allowed for this, and could he negotiate with the companies to offer him a 'bulk-use' concession to reduce the costs? He would need, too, to develop a method of coding the readings for transmission, squeezing the most information into the fewest characters, while – since telegraphs were notoriously prone to mis-transmission – maintaining some sort of security to ensure that a complete and accurate message had been received.

And that was only half the story. How was he to communicate his warnings to the remote coasts where they were most needed? Who would receive them? How would they then transmit them on to the seamen who must be warned? It was obvious that he would need to enlist the cooperation of some agency with stations around the coast, and the obvious agency to call upon was the Coastguard. And since the Coastguard was then controlled by the Admiralty, that meant exerting his influence at Admiralty House to win it to his cause. The final link in the chain, the signals to seamen, would have to be visual. Something that could be hoisted on shore, large enough to be seen at sea, robust enough to withstand rain and gales, simple enough to be understood instantly, yet capable of carrying the vital information of force and direction of approach.

By coincidence, in another of his guises, FitzRoy had been collaborating with Sir John Herschel in a project that called for a similar type of warning signal – the matter of lighthouses and how to improve their visibility at sea. After several rounds of tetchy spatting between the two of them, a signal had evolved that FitzRoy now adopted as 'the FitzRoy cone' – an object that, like the FitzRoy barometer, would take his name forward into history. It was an ingenious device. Hoisted on a mast, a cone looked like a triangle from the distance and did so from whatever angle it was seen. With its point up for north or down for south, it could indicate the approximate direction from which the storm would appear. A cone on its own would signify a cautionary warning, but if a drum, which from the distance resembled a square, was hoisted also, then the signal indicated imminent danger. The cone and drum could be cheaply made and light – FitzRoy suggested collapsible versions constructed from hoops and black canvas, instantly available yet easy to store. With these devices, the telegraphed storm warning need only be short, 'Caution North' for example, or 'Danger South'.

By the beginning of April, FitzRoy was ready to submit a detailed plan of how to proceed to James Booth. The plans were sound and his confidence must have been high. Deflation was to come like the prick of a pin to a balloon – the weapon seemingly innocuous, the result catastrophic. The 'pin' was a report sent originally to Professor Airy, the Astromoner Royal, which Airy duly passed on to FitzRoy. The date he received it, judging by when he responded to it, was 13 April – ominously, a Friday. The report, in French, was from Le Verrier in Paris, and it contained proposals for a Port Meteorological Telegraph Service.

Despite the Imperial decree of 1855, and its hopes of a future storm warning system, the only progress made by Le Verrier in the intervening five years had been his network of telegraph operators and his daily published weather reports. There were, as yet, no attempts to transmit specific storm warnings – people were simply left to interpret the weather reports for themselves. Now Le Verrier intended to rectify this and progress to the next stage. Initially, following Liais's examination of the charts of the Balaclava storm, both he and Liais had been excited by the discovery of what appeared to be waves of undulating atmospheric pressure accompanying the storm and they had hoped this would lead the way to a mathematical formula from which a storm could be predicted from measurements of pressure. So far, no such magical formula had been forthcoming. The alternative was much more prosaic, but had the attraction of being technically achievable. As soon as a storm was notified as having arrived somewhere in Europe, the plan was to follow its progress and notify by telegraph all ports that were likely to be in its path. Although modest, the scheme still required effort and funds over and above the current extent of the telegraphed system, and to operate to its full potential, it also needed the cooperation of other countries, since, if confined to France, the number of storms detected early enough would be far fewer. Britain's participation was particularly imperative, as most storms appeared to originate out in the Atlantic, hitting Ireland first. Le Verrier, therefore, sent a copy of his report and a letter requesting a telegraphed exchange of observations to Professor Airy, who, being the astronomer in charge of the Greenwich Observatory, was the man he was accustomed to dealing with in England. Airy recognised that this more rightly came into the sphere of the Meteorological Office and immediately passed the report and the request on to FitzRoy.

FitzRoy was horrified. It sounded suspiciously as if Le Verrier was on the point of installing a system of telegraphed storm warnings before he had had a chance to complete his own. It was probably of no relief, either, to read Le Verrier's detailed explanation of his plans, which mirrored the original proposal by Locke to the British Association, operating merely as a messaging service.

FitzRoy could accept that Le Verrier's system made some sense for the French. France, situated on a large land mass and far broader across its east–west beam than was Britain, stood to gain much more from a purely factual warning network than Britain did. Storms usually rolled in from across the Atlantic, and those that hit France first crossed over the Bay of Biscay, battering the French west coast before travelling on over Europe. Information of storms approaching over Ireland would be an enormous advantage to the French, but reciprocal information from the French would be of restricted use to FitzRoy. The benefits appeared to be loaded in the Frenchman's favour.

Fearful of finishing an ignominious runner-up to the French, FitzRoy slaved at his desk, completing his implementation plan. In defending the possible shortcomings of his scheme, he wrote:

> since meteorology is surrounded with uncertainties requiring practical and theoretical qualifications as well as aptitude in those who would foretell rather than only chronicle the weather, it is not expected that accuracy in all cases will be attainable – but sufficient accuracy has already been proved to justify the expense of an experiment.

Unfortunately for FitzRoy, M. Le Verrier was also occupied in putting pen to paper. News of FitzRoy's impending predictive storm-warning scheme had reached him in Paris, and he, too, was loath to be beaten into second place, especially by an English admiral with delusions of being a scientist. He was envious, too, of FitzRoy's daring in attempting to use a scientific approach to predicting storms, rather than relying on communication of an actual storm. It was, after all, where in his own heart he would much prefer to go. FitzRoy had to be dissuaded. What he knew, and FitzRoy did not, was that even his modest scheme for France had not yet been approved by the French government. Funds, and therefore storm warnings, were not imminently forthcoming. If FitzRoy went ahead now, it would be he, not Le Verrier, who would lauded as King of the Storms – and the French government might well opt for the cheap option of taking British warnings rather than expensively generating their own. Carefully, he composed his words.

His letter arrived in FitzRoy's hands at about the same time as FitzRoy's own proposals reached Secretary Thomas Farrer. As FitzRoy ploughed through the letter, digesting Le Verrier's French, his heart must have collided with his ribs when he read: 'Si donc vous me le permettez, Monsieur l'Amiral, j'oserais vous recommender de ne pas repousser ce que nous proposons en s'appuyant sur ce qu'on pourrait faire d'avantage.' Le Verrier was, in effect, warning him off – do not go ahead with your prediction service, for you will undoubtedly compromise the better scheme that we are trying to achieve.

The implications of this statement were frightening for FitzRoy. He would have been able to visualise his superiors looking more quizzically on his own proposals – why go for the riskier predictions if the French are convinced that a system that solely transmits facts does actually work? The damage Le Verrier's views could inflict did not bear thinking of.

Unfortunately, his worst fears were quickly realised. As soon as Farrer got wind of Le Verrier's warning, he rushed around to his boss – James Booth, Joint Permanent Secretary – and the two of them called in FitzRoy. They wanted, they said, to question him more fully before submitting his plan to the President. In fact, it was obvious that they were completely won over by Le Verrier's argument. He 'seems to deprecate, till the system is fairly started, any attempt to foretell weather' Farrer told FitzRoy, and accused him of proposing a system far more complex than Le Verrier's, without explaining the scientific basis behind it.

He was right; FitzRoy had not provided a scientific explanation. It might seem arrogant on FitzRoy's part not to have attempted to explain his scientific reasoning, and there might well have been an element of intellectual superiority in his attitude. However, by the time he had reached the stage of being invited by William Cowper to submit his detailed proposals, he could reasonably assume that the principle behind his scheme had already been accepted – and, but for Le Verrier, this would quite probably have remained the case. It does not mean the principle was understood.

It seems fairly obvious to us today. We are accustomed to televised, printed and Internet weather maps, computerised simulations, satellite pictures and all the paraphernalia of modern technology that allows us to view the stream of depressions scurrying eastwards across the Atlantic towards Europe. We understand that, like contours on a map, the closer the isobars, the deeper the depression, and it does not really take a forecaster standing alongside to tell us when the outlook is stormy. We can easily comprehend, too, from the way the wind swirls around a depression, that, just because the wind is from the north-east, it does not necessarily mean that the weather is approaching from that same direction. In fact, in Western Europe it very rarely does approach from anywhere other than the west. We also realise that some depressions follow a more northerly line than others – some hit land in Scotland, or Northern Ireland, or Cornwall, and some miss Britain altogether to vent their wrath on France or Spain. However, even to the educated and intelligent of the mid-nineteenth century, there was nothing instinctive in this. They had never seen a weather map, let alone a satellite picture. Some, such as the sailors, fishermen and farmers who depended upon the weather, were more aware of its vagaries, watched the clouds, knew what the different wind directions might bring, but for most town-dwellers the weather was an arbitrary, capricious affair – the wind blew, the sun shone, the rain came or a storm hit and that was fate.

To Thomas Farrer and James Booth, as career civil servants, it would have seemed more logical to trust factual evidence – that someone had seen with their own eyes a storm over Land's End or over the Yorkshire coast – than to believe in the ability of a naval officer, admiral or no, to consult the oracle of a few scattered barometer readings and predict the coming weather. When the French equivalent of the Astronomer Royal added weight to their conviction, it was going to take a very strong argument indeed to convince them otherwise.

FitzRoy tried. Since they had asked for a scientific explanation, he gave it to them in a flood of highly technical explanations, spattered with extensive quotes from his own papers, Dové and the works of other eminent meteorologists. In a hastily compiled outpouring, his desperation boiled to the surface – how much this experiment meant to him, his dread that Le Verrier's meddling would block him from his goal and his sheer frustration that no one else could understand why the French system would not work in Britain and why his proposed scheme would. He tried attacking Le Verrier. He claimed, with some accuracy, that the only reason Le Verrier was not proposing to make predictions was because he did not have the means to do so. 'In this office,' said FitzRoy, 'information has been accumulated and digested for five years with the object, among others, always in view of foretelling the weather.' This 'object' was probably news to Farrer and Booth, both of whom would have been happily assuming that their Meteorological Officer took as his top priority the tabulation of figures, and nothing more. They were vehemently disillusioned of any such assumption, as, with some heavyweight name dropping, he reinforced his intention not to continue to pour resources into statistics. Quoting Sir John Herschel as saying 'a man might as well register his dreams as record details from which no results could be derived', he declared it was a much better use of resources to divert them to the much more publicly rewarding work of weather prediction, however approximate, 'and such an approximation may save agriculturists' crops, ships and even lives'.

Storm warnings were an important means to saving lives, but they were also vital to saving FitzRoy's reputation. So vital that he was prepared to state officially his intention of steering his effort away from the steady accumulation of meteorological statistics and onto the uncertain course of weather prediction, a major turning point in the work and purpose of the Meteorological Department. FitzRoy, the skilled commander, did not take risks with the safety of his ship, unless circumstances dictated. Circumstances here were dictating.

However, at that moment, far from making a significant departure, FitzRoy and his plans looked like being scuppered before they had nosed away from the quay. James Booth, for one, could not make any sense of what FitzRoy was trying to say. In fact, he was so confused he seriously thought this was all something totally different and that FitzRoy proposed accomplishing the objective of the British Association by a separate scheme altogether.

As a result, FitzRoy suspended his efforts at trying to convince and composed a straightforward précis of the operation of his scheme to be used as the basis for submission to the Board of Trade President. In its 'no frills, no explanation' tone can be sensed a resignation to his fate. From this, Farrer summarised the proposed scheme, stating its dual purpose as being to collect the facts and, from them, make conclusions about the coming weather. Then, knowing exactly the effect it would have, he added Le Verrier's warning. Indeed, the weight of Le Verrier's pronouncement did set the seal on FitzRoy's fate. The President proclaimed that telegraphic transmission of facts about the weather was perfectly acceptable, but the government would not take responsibility for drawing conclusions from these facts nor for the issuing of predictions.

FitzRoy must have reeled at the blow. He had, however, been left a loophole. While the government would not take official responsibility for FitzRoy's predictions, Milner Gibson had decreed that the British Association was at liberty to use government-supplied data as it wished, and, since FitzRoy was a member of the Association, then he was free to issue all the predicted warnings he wanted, provided he accepted responsibility for their accuracy. He could look at this let-out as being a gift from Milner Gibson – you make it work, as you say you can, and the glory will be all yours. FitzRoy determined that it would work.

With summer and the promise of a sunny future warming his soul, Admiral FitzRoy set out for Paddington a couple of weeks later to take the train to Oxford, for the 1860 General Meeting of the British Association. As he chose his seat and the porter swung his baggage onto the shelf above, he must have smiled to himself, thinking of the paper packed in his case – a paper he had prepared on British storms that he was to present to the meeting. More vitally, this was the vehicle he had chosen for the launch of his new storm-warning system. Only ten months on from Aberdeen, he could be smugly satisfied that he had achieved the remarkably speedy success of bringing the association's resolution on Telegraphed Storm

Warnings before the Board of Trade and having it signed for action. He would make his big announcement in front of the assembled luminaries, hopefully to stun them all with the audacity of his progressive scheme.

There would have been nothing to hint at trouble ahead when he strode across Oxford on Thursday, 28 June, the day before his own performance was scheduled. Then, as now, the bustle of the modern-day city would be chopped off instantly at the College gatehouse. Within lay ancient walls and lofty halls accustomed to a tranquility broken only by the clash of intellectual debate. In such a setting that afternoon Dr Daubeny of Oxford would take to the lectern to present his paper on the Sexuality of Plants. The subject was seemingly innocuous and so far from FitzRoy's own field to offer apparently no threat to him – yet it could, for Daubeny's paper came primed with the additional phrase 'with particular reference to Mr Darwin's work'. FitzRoy would have joined a room drumming with anticipation, eager to hear what the ensuing debate would bring. An animated exchange did indeed develop, in which the idea of man's descent from the monkey was drowned out in guffaws. It was no doubt with a wry smile of triumph that FitzRoy went to his supper that evening.

On Friday, in contrast with the controversies of the day before, he delivered his paper in an atmosphere of calm, his performance polished, his theories well rehearsed. He gave a comprehensive survey of storm theory as it was known and described, with detailed diagrams, the findings of the *Royal Charter* storm and the patterns of weather behaviour learnt from it. Finally, he would have turned from his diagrams, stepped up once more to the lectern, paused while turning to the next sheet of his script, and then looking straight at his audience would have announced:

The British Association has made application to Her Majesty's Government to authorize arrangements for communicating warning of storms from one part of the country to the other; and in conclusion I will read the details of that arrangement which promises to be beneficial . . .

Authority being thus given to collect and communicate, by the telegraph, particular meteorological intelligence, a commencement may be made on the 1st of September, as the plan proposed is simple, and the machinery is ready. Once a day, at about nine a.m., barometer and thermometer heights, state of weather, and direction of wind will be telegraphed to London, from the most distant ends of our longest wires, – namely, Aberdeen, Berwick, Hull, Yarmouth, Dover, Portsmouth, Jersey, Plymouth, Penzance, Cork, Galway, Londonderry and Greenock. Facts sent thus from five of these places, will be put into one telegram and sent to Paris immediately, when a corresponding communication will be made from the southward Atlantic coasts. When threatening signs are not apparent, no further notice will be transmitted to or from London on that day, respecting weather. But when indications are such as to warrant some cautionary signal at a certain part of, or along all our coasts, the words 'Caution, – North' (or 'South') will be sent to some of the thirteen places specified, or to all of them, on receipt of which a cone will be hoisted at a staff . . . By vigilance at the central station, and by taking great care to avoid signalling too frequently, much may be done towards diminishing the losses of life on our increasingly crowded coasts.

It was a splendid performance, detailed and confident. Whether FitzRoy himself felt quite so confident is open to doubt. The whole system depended on him. No one else had the skills to undertake it. He had laid down a hefty burden on himself, for if he failed to warn correctly, then the whole system would be useless. But then this was a man who had lived his life reliant on his own abilities and the strength of his beliefs.

The very next day was to bring one of the greatest challenges to those beliefs he had ever faced.

The Saturday lecture was to be read by the American, Dr Draper, concerning the Intellectual Development of Europe, and it too, like Daubeny's Thursday effort, was to be discussed in the light of Darwin's theory. After the controversy of Thursday, the debate promised to be lively. Entertainment on this scale was rare outside term time, and so many people flocked to the lecture that it had to be hurriedly transferred from the small lecture room to the museum library, where several hundred people crammed into the stuffy heat of the summer's afternoon. Among them was Robert Fitzroy.

Lined up for the protagonists were two stalwart supporters and friends of Darwin, Thomas Huxley and the botanist Joseph Hooker. The Bishop of Oxford, Bishop Wilberforce, son of the emancipator of slaves and well-known 'creationist', was set to make the reply. The bishop was famed for his vigorous, lucid yet soft-spoken manner, and to those who opposed Darwin's views, his speech must have come to their ears like an oasis of sanity amidst a desert of profanity. There came from the floor so many demands to speak that the privilege was restricted to those who possessed a scientific reason to comment on the paper itself. FitzRoy, however, was given his chance. His role on the *Beagle* gave him credence, and perhaps his opinion carried sufficient weight for the assembly to calm to hear him out. Exactly what he said is uncertain. The *Athenaeum*, which published a sanitised version of the event, merely reports him as saying that he 'regretted the publication of Mr Darwin's book and denied Professor Huxley's statement that it was a logical arrangement of the facts'. What he actually said must have been an attack of some vehemence in support of the literal meaning of Genesis, for it provoked Darwin to write in a letter to Henslow a couple of weeks later 'I think his mind is often on the verge of insanity.'

Darwin's comment on his sanity is a perspicacious observation in view of the events that were to unfold, and contains, even at this stage, an element of truth. There were great battles going on within the FitzRoy psyche, and the fight for fundamentalism was only one. His seeking of fame, his aspiration to be accepted as a scientist among scientists in the field of meteorology and his heartfelt philanthropy, which urged him to attempt to ease the lot of those who put themselves daily at risk in the elements; all these drove him on. And it was now, in the summer of 1860, that they were all tantalisingly in reach. The moment was in sight when his connection with Darwin would be exorcised by the success and acclaim of his own original project, his storm predictions.

From his speech, the members of the British Association might have been forgiven for expecting that all was cut and dried, that the machinery for implementing the 'simple' plan was in place, and that, come 1 September, messages would be whizzing round the country, warnings would be delivered when and where they were needed, and never again would a sailor set out to sea to face the severity of an unexpected storm. FitzRoy himself believed that all was prepared. His head of steam was up and his ship was pulling at its anchors, ready for the off. But departure was not imminent. The processes of government administration did not run as quickly or as smoothly as FitzRoy gamely expected.

Forecast

Black cloud shrouded the dawn, and the gale swept sheets of icy rain across the cottages huddled along Blakeney quay. Beyond the marsh flats, the sea pounded in a deafening roar. None but the foolhardy would venture out in such weather, yet a boat crewed by men at oar was clawing its way along the channel towards the sea. A distress signal had been seen around the point, and Blakeney beachmen would never shirk their duty in attempting to save the crew of a sinking vessel. Rounding the point, the small boat was hit broadside by a mountain of a wave and capsized. Nine men lost their lives, including three brothers, John, Samuel and Thomas Johnson. As the morning light grew, the news of the deaths cast the village into a deep gloom.

The crew the men had died trying to save belonged to the barque *Favourite*, on its way from Hartlepool to Torre del Mare in Spain with a cargo of coal. It had been blown onshore at Warham Hole near Blakeney in Norfolk in the vicious north-easterly gale of 9 February 1861. Fortunately for the crew, they were rescued by another boat, as were the crew of the *Kingston*, a schooner also out of Hartlepool on its way to Plymouth and grounded by the fierce seas on the beach at Blakeney. Further north, off Hartlepool Bay, the devastation was much greater. A fleet of ships had set out from the Tyne and the Tees in an effort to beat the storm, but the gales caught them far too near in to tack out to sea. Every ship was forced to turn and run before the wind, hoping to reach safe harbour. Thousands of people flocked to the cliffs and beaches, as the heavy laden vessels were blown relentlessly towards the shore. The crowds could only watch as ship after ship beached on the sands or crashed in against the rocks, smashed to splinters, their crews screaming into the gale as seas swallowed them. In the course of five hours, fifty vessels came to grief along the coast from the Tyne to

Whitby, of which thirty were total wrecks. Thanks to the heroic efforts of four lifeboat crews, many lives were saved that day, but at least eight ships foundered with the loss of all on board.

In *The Times* on 12 February, alongside gruesome reports of the wrecks, there appeared a letter from Robert FitzRoy.

A column of your paper is filled today with accounts of a severe north-east gale. All the much frequented parts of our coast might have been warned – a very few places actually were warned – three days before this storm. On Wednesday last [7 February] the following notice was given at Aberdeen, Hull, Yarmouth, Dover, Liverpool, Queenstown, Valentia and Galway (besides other places), by telegraph:–

Caution – Gale threatening from south-west then northward. Show signal drum

This was Robert FitzRoy's first telegraphed storm warning. The signal to show the drum meant not just that gales were on their way, but extreme danger was also indicated, and he was quite correct. Early February that year had been extremely windy and very wet. From the 4th to the 9th nearly 2 inches of rain had fallen in some parts, accompanied by thunderstorms, hail and snow. Teesdale was covered in several inches of snow and the Vale of Pickering in Yorkshire was inundated with its fourth flood of the winter. The wind had been blowing steadily from the south-west and was still doing so on the Wednesday. FitzRoy, though, could foresee change and issued his warning. On the morning of Thursday, 8 February, the wind suddenly veered, sweeping round through westerly to northerly, to become, in the early hours of the 9th, the north-east gale that had been so destructive of shipping on the east coasts.

So why, despite the warning, had that large fleet set out from the Tyne and the Tees into the teeth of an imminent storm? One explanation was that the barometer had been rising for a couple of days, and accepted wisdom was that the barometer always fell when a storm was imminent. However, as FitzRoy pointed out in a PS to his *Times* letter:

I would repeat what has been reiterated and explained elsewhere [i.e. on his barometers and in his barometer manual] – that the air is lighter as well as warmer during southerly winds; heavier and colder before and during northerly; and that the influence of either is shown by instruments some hours, if not days, before actual alteration is visible to ordinary notice.

Someone in the Tyne and Teesside ports should have read the barometer and interpreted it correctly – and perhaps someone did, for another report in that morning's *Times* claimed ships had risked sailing while they could, since it was feared the sea would rise even higher, tying cargoes to port for several days. Mercenary considerations outweighed the risk to life, so the women and children of Blakeney had to lose the brave men who were their sons, husbands and fathers, and all in vain. It was the very scenario FitzRoy had hoped to avoid.

On the face of it, it may seem that the Board of Trade's telegraphed warning had been blatantly ignored, but that was not the case. This first warning, though its coverage appears extensive, was not signalled at any of the ports from which the doomed ships sailed. Despite eight months of preparation and organisation, despite the satisfactory telegraphing of weather reports across country and over to France since the previous September, when FitzRoy telegraphed his first storm warning that February the mechanism for raising warning signals was still woefully inadequate.

The problem was not with the telegraph. One of FitzRoy's first tasks had been to persuade the commercial telegraph companies to fit in with his plans, and they had responded well, happily cooperating in sending their readings from instruments provided by him. Next, he had approached the Admiralty for its permission to allow the Coastguard to watch for the signals at the telegraph stations and relay the warnings along the coast. The Lords of the Admiralty were 'disposed' to put his scheme into operation, but wanted to make sure it was the Board of Trade and not they who would be footing the bill for all these expensive telegrams. FitzRoy duly reassured them, and flattered himself that nothing now stood in his way.

With that accomplished, and with new staff in place – including two young and fit messengers to run back and forth to the London telegraph office – the first telegraphic communications were successfully exchanged with France, as prescribed, on 1 September. Two days later, the first of a daily set of weather reports (facts of weather past, not weather to come, not yet awhile) appeared in various daily newspapers and was posted at Lloyds. All that remained was to supervise the manufacture of warning cones and drums and to distribute them to the designated telegraph offices and Coastguard stations.

Opposition to his scheme was inevitable. Influential scientists needed to be co-opted to his point of view. Professor Airy at Greenwich was a prime target. 'As a matter of Abstract Science, you think Meteorology desperate,' he declared to Airy, 'but as a matter of practical national use in the proper hands, its value may possibly be great . . . Do not atmospheric indications and laws seem to be even now reducible to almost systematic expression, not, however, as may be desired, mathematical?' Therein lay the nugget of scientific criticism – to Airy as an astronomer and to many other followers of 'abstract science', mathematics was the only possible manner of systematic expression. The professor could not have been too taken aback, though, since a couple of weeks later he invited FitzRoy to add to the impressive list of learned societies who had accepted him, by inviting him to join the Royal Astronomical Society.

Sir John Herschel, on the other hand, often a cantankerous critic of FitzRoy's, was for once on his side, backing his telegraphing of winds and cyclones as the most valuable meteorological information that could be telegraphed.

A little earlier, in October, there had been a flurry of excitement when Matthew Maury arrived in London. He came in order to copyright the British edition of his *Physical Geography of the Sea*, which he wished to dedicate to Lord Wrottesley in recognition of his contributions to the cause of meteorology. Whether he and FitzRoy actually met is not recorded, but meteorological circles, especially those frequented by Wrottesley and the gentlemen of the British Association, were small and exclusive, so it would have been difficult for them to avoid one another had they tried. Not everyone fawned over Maury, though. His popular acclaim appeared to garner envy from his British counterparts. Even FitzRoy had remarked, 'as I am aware he occasionally theorises when he has not facts enough for philosophy', while in Sir John Herschel, who rarely had a good word to say for Maury, he sparked a stream of spiteful criticism. 'I wish his language were less inflated and his dynamics sounder,' he remarked to FitzRoy and went on to refute Maury's notion that sea currents were caused by changes in the density of water.

Maury's freewheeling around London was brought to an abrupt halt by an item of devastating news brought in early December on a ship from New York – the secession of South Carolina from the Union. This, said Maury, was the worst of news, for it could only pit friend against friend. Remaining neutral was not an option, for the place of his birth dictated his allegiance. Maury, despite his current home in Washington, was Virginia born and Tennessee raised, and where South Carolina went the other Southern states were bound to follow. He withdrew from polite London society and headed home to certain war. FitzRoy must have viewed his going with regret, for even at this early stage of events, it was obvious that Maury would be lost to meteorology, and therefore to FitzRoy. Despite FitzRoy's criticism of the man, Maury had remained to him a good friend and a true supporter.

It was a tense time, waiting while the machinery for warnings was engineered into place, and FitzRoy was not alone in finding it so. Impatience broke out in the most unlikely places, such as at the Sailors' Home in Great Yarmouth on the Norfolk coast. FitzRoy had had trouble with these keen old tars before, when they had applied to be a warning station in the hope of obtaining some free instruments for themselves. He had turned down their application, but nothing daunted, they had taken it upon themselves to begin hoisting their own warnings, based on the published weather reports. No doubt FitzRoy sympathised, but he was forced to step in to curb their enthusiasm when they began to boast publicly that their homespun signals were sanctioned by the Board of Trade. Wait a few days, he told them, and the proper signals will start to appear.

The few days dragged on to a few weeks. Christmas came and went. In his end-of-year report to the President of the Board of Trade, FitzRoy complained that, although the telegraphic system was in operation and the machinery for warnings was ready, and although the Admiralty was sounding out the Coastguard, no cautionary signals had yet been made, because they had not yet been authorised. The authorisation he was looking for was monetary, and it came in January in the form of a £500 appropriation for a one-year experiment, backdated to the September commencement of telegraphed messaging.

FitzRoy dashed off the news to the telegraph companies, asking them to stand prepared to hoist storm warnings, and also to the Controller of the Coastguard. The telegraph companies gladly cooperated, and almost immediately, in time for that February storm, his very first warning was issued and the first cautionary signals hoisted. However, to his dismay, the men of the Coastguard knew nothing of what was expected of them. FitzRoy had fondly imagined that the Admiralty staff would have made contact on his behalf, but this had not happened. There is a contained rage behind his letter to the Admiralty asking that 'a certain degree of official notice be taken of these cautionary signals at ports and along coasts by HM officers in Command; as well as by the Coast Guard, generally (already approved by their Lordships)'.

His reprimand had an instant effect. The Secretary to the Admiralty, Romaine, requested that patterns for the signals be sent to Portsmouth and Devonport dockyards, so they could begin manufacture. Unfortunately, it was not FitzRoy to whom he addressed the request, but Board of Trade Joint Secretary Booth, and Booth was not cooperating. These signals are nothing to do with the Board of Trade, he informed the bewildered Romaine; we have not sanctioned them. Ask FitzRoy; this is his scheme. It could have ended there. The Admiralty might have blocked FitzRoy's precious experiment by refusing Coastguard compliance. But here, at least, FitzRoy's rank and reputation counted in his favour, and the Admiralty was willing to assist him, provided, of course, he would take all the blame if it went hopelessly wrong. FitzRoy had already taken on that responsibility, and anyway the experiment simply could not fail.

The trouble with such an exposed experiment was that it was open to constant public scrutiny, and the public soon let him know what they thought. He could be man enough to concede when he was genuinely wrong, but unfortunately the public, and the press, were not always very accurate in keeping to the facts. Not long after his first storm warning, *The Times* claimed a troopship had set out from Woolwich despite a prognostication by Admiral FitzRoy, received there from Lloyds, of an approaching gale. The officers apparently 'expressed no inclination to proceed' but sailed anyway. FitzRoy was incensed.

I feel it a duty to lose no time in saying that no such 'prognostications' were made by me, that 'Lloyds' did not send such to Woolwich or elsewhere and that officers in Her Majesty's services do not hesitate to obey lawful orders, rather than their own inclination.

If you want to prove you are right, you have got to show you are right, because the world is full of folk quite willing to prove you wrong. The public launched into the job with gusto, amateur observers everywhere delighted at this sudden apparition of a target at which they could aim the results of their work. They were soon sending him their own interpretation of whether his warnings matched the weather.

FitzRoy retaliated. The facts proved, said he, that, the Hartlepool disaster aside, few wrecks occurred during the tempestuous weather of February and March that year, and comparatively few since. And he quoted from *The Times* to prove his point:

Not withstanding the extremely stormy weather that has prevailed in the North Sea during the first three months of 1862, the coasts of Durham and Northumberland have had a fortunate immunity from shipwreck, there having been no serious shipping disasters upon that line of coast since November 1861.

FitzRoy expected detractors, but the approval he sought was from the people who truly mattered, the seafaring community. What did the sailors think? Were the masters of small coasters happy to take heed of his warnings? Were there fishermen out there who blessed him for saving their lives? The best way to find out was to ask them, so a questionnaire was circulated to fifty-six different coastal authorities to canvass their opinion. As the sheets came in, FitzRoy's smile began to broaden, and, when all fifty-six were back in his hands, the final results were an overwhelming endorsement of his scheme. Forty-six were decidedly favourable, nine approved with some reservations and three – only a tiny minority – were adversely critical.

His critics could be wordy. William Maclean wrote from Great Yarmouth with a long list of objections. Maclean was of the old school, and appeared to view the scheme as namby-pamby mollycoddling, which would only make mariners timid. In some instances, he noted, fishermen had refused to sail, yet no storm came within twenty-four hours and they had sat in port needlessly, losing much time and profit. In other cases, the signals could be downright misleading.

A vessel whose destination might be reached in 20 hours might make her voyage, while delay would be the cause of a storm overtaking her. This actually happened to two vessels which were ready to sail from this [*sic*] during last autumn. The one that pushed on notwithstanding the storm signal being up reached her destination in safety; the vessel which delayed to sail and put to sea afterwards was caught in the storm and was lost.

Maclean's views were reflected by the Port of Shields Authority, which declared that the warnings 'are certainly not trusted by seafaring men . . . They are not found useful, but have often caused unnecessary delay and inconvenience, both to owners and seamen without any corresponding good result.' And the men of Cowes agreed: 'wedded to notions they have gained by long experience in watching local indications of weather', they preferred to trust to their own judgement.

FitzRoy retorted to these critics that perhaps they should counterbalance the cost of delays in port with the savings that had been made by not losing vessels in a gale. As for trusting one's own experience, then he was all for it – he had always claimed that the warnings were merely cautions to be on guard. 'Notice your [weather] glasses and the signs of the weather,' he said, and decide for yourself.

Never mind the sourpusses and the brickbats, approval outweighed them, and it came in heart-warming stories, such as from the gentleman wishing to travel to Ireland with an invalid lady. The weather had been fine in London, but he had heeded the admiral's warning and decided not to set out. That night it blew a hurricane to the west of Ireland and a gale in the Irish Sea. A fellow admiral, Admiral Evans, was particularly impressed by a certain warning issued at Liverpool, and passed on to FitzRoy the report from George Hamlin, the Harbour Master.

Yesterday about 3 p.m. we were warned by telegram of an impending gale and about high water (8.40 p.m.), it burst upon us like a thunderclap . . . I took the precaution to circulate the warning as widely as possible, which was the means, perhaps, of preventing some accidents.

But it was probably the short notes like the one that arrived from Devonport that warmed FitzRoy the most. It said: 'Seafaring folk are proverbially averse to novelties but already some of the oldest and most obstinate begin to look with some degree of confidence to the display of the cone and drum.' That was the very justification he craved, and to receive it made the effort worthwhile. Charles Darwin might well be stirring scientific excitement and gaining all the accolades, but he was not – and could not – gain the hearts and thanks of the ordinary people.

After the public brouhaha that accompanied the birth of storm warnings, forecasts arrived at the British breakfast table almost unnoticed. FitzRoy casually slipped the first of them into the weather report that was printed in the newspaper on Thursday, 1 August 1861.

General weather probable during the next two days in the –
North   Moderate westerly wind ; fine
West   Moderate south westerly wind ; fine
South   Fresh westerly ; fine.

Soon his forecasts were being published six days a week, in six daily papers, one weekly paper and at Lloyds, the Admiralty, Horse Guards and, of course, the Board of Trade. There was no great battle for the approval of his masters for this, since there was no new system to set up and no extra expense. Weather forecasts were born out of storm warnings, and their future had been secured on the day that Milner Gibson had allowed him to use government-obtained data as he wished. To predict a storm, he needed to predict the weather. If the weather was not to be stormy, then the prediction he had made had no further use – it might as well be discarded into the nearest waste bin. But why do that? Why not give the public the benefit? As FitzRoy said himself, 'forecasts add almost nothing to the pecuniary expense of the system, while their usefulness is more and more recognised. Storm warnings arise out of them, and their negative evidence is no less valuable to a person about to make a crossing.'

He was not the first to make public weather predictions. Joseph Henry, perhaps getting wind of FitzRoy's telegraphed prediction activities, took pains to explain the Smithsonian Institution's weather-reporting achievements.

> Our system of telegraphing the weather during the last winter was a source of great interest . . . When our reports are full, particularly from the west [of Washington], we scarcely ever fail to foretell, for nearly a day in advance, the state of the weather. Various suggestions have been made . . . to exhibit daily, by maps, useful facts regarding weather. But in this country the expense of making, transmitting and publishing soon enough to be valuable has impeded actual organisation of any system better than that established by the Board of Trade.

In Holland, Christoph Buys Ballot had been attempting estimations of the coming weather using his rules of differences in barometric pressure. Like FitzRoy's, these estimations were based on prediction rather than the tracking of known weather. Again, like FitzRoy, Buys Ballot needed to construct a prediction every day, in order to recognise when to issue a storm warning. He had, though, restricted his published output to those warnings, not daily forecasts, and although his methods were different, the results bore comparison with FitzRoy, as Buys Ballot explained himself to the British Association in 1963:

> on the 1st of June 1860, the first telegraphic warning by order of the Department of the Interior was given in Holland . . . All of you know how amply Admiral FitzRoy has arranged the telegraphic warnings all over England. The rules over Holland have answered well . . . it appears I have warned from my four stations just as Admiral FitzRoy has done from his twenty.

Buys Ballot's inaugural warning preceded FitzRoy's by eight months, making him the first to issue a storm notice based on weather prediction. But Robert FitzRoy was indisputably the first with a general weather forecast, since 'forecast' was the term he himself invented. His storm warnings he called 'cautionary', but 'cautionary' could not be applied to day-to-day weather. Where was the threat in a clear, sunny day? So what was his definition of a forecast? 'Prophesies and predictions they are not:– the term "forecast" is strictly applicable to such an opinion as is the result of a scientific combination and calculation . . . They are . . . deductions . . . for which rules have been given.'

The Meteorological Statist had gradually and unerringly transformed himself into Weather Forecaster. Numerical tabulations had lost any fascination they might once have held. His department still struggled on, trying to accumulate more and more observations from ship and shore, but he could not see the point. 'Stones may be shaped, bricks may be accumulated, but without an object in view – without an edifice to be constructed – how wearily unrewarding to the mind would be such toil, however animated . . . by true scientific faith in future results.'

Shelves full of dusty ledgers might be useless piles of bricks to FitzRoy, but someone, somewhere, had an edifice awaiting construction and was desperate for those bricks. And since the Meteorological Office was unwilling to supply them, he was forced to source his building materials elsewhere. In July 1861 a circular letter arrived at the doorsteps of meteorological observers all over Europe.

The letter was addressed from 42 Rutland Gate, London, and was signed 'Francis Galton'. Thwarted by FitzRoy, and not for the first time either, Galton was not a happy man.

S I X T E E N

Meteorographica

As a basis to future efforts, I here invite Meteorologists who have
been in the habit of contributing observations to any Society, and
are therefore familiar with methods of observing, to co-operate
with me during the whole of next December, in order to obtain a
series of aerial charts of Northern Europe . . . A copy of these will
be presented and forwarded, by book post, gratuitously to every
Contributor who will send me, postage free, a series of reduced
observations and other information, according to the subjoined
conditions.

   The result of a wide system of co-operation such as I propose
will be the accomplishment of a valuable piece of scientific
work, that will also help to afford an answer to the question
whether synchronous charts may hereafter be printed regularly,
with success.

Headed 'Synchronous Weather Charts', and accompanied by a
sample chart of England for January 1861, this circular had
come completely out of the blue. Apart from an educated
philosopher's general curiosity, Galton had not previously
shown a predilection for meteorology, so, when this circular
fell into his hands, FitzRoy was no doubt both surprised and
concerned. He had already been forced by Charles Darwin into
a theological argument he would rather have done without.
Now the cousin, Francis, stood ready to rush in and tread all
over his precious meteorological preserve.

   However, the true surprise was that Galton had not been in
contact with FitzRoy before, on the subject of the heliostat, his
sun-powered signalling invention that he had submitted to
FitzRoy. Since the diagrams had disappeared into the
Admiralty, now nearly three years ago, there had not been a
peep of a response – and FitzRoy was quite happy to leave it
that way. But that little instrument, nifty as it was, was merely

one of those minor irritations that crept under his skin every time a Darwin crossed his path. This newest incursion was much more serious. It smacked not of skirmish, but of invasion, and constituted a very real threat.

Within the interlinked chains of Victorian philosophic circles, it was likely that Galton and FitzRoy had met from time to time. They were both Fellows of the Royal Society, and members of the Royal Geographical Society – FitzRoy on account of his surveying and Galton for his foreign explorations and books of his travels. Up until now, the multitalented Galton had focused on geography, specifically on the practical and useful applications of the subject, and one of his unrecognised achievements is as the writer of a series of popular travel guides, *Vacation Tourists and Notes of Travel*, which must be among the earliest thoroughly practical tourist guides in the modern form. Of course, one item of information vital to the comfort and safety of the tourist to a foreign country is a knowledge of the climate. And that was where the link with meteorology came in.

By the mid-nineteenth century, the world was opening up and travel to exotic locations was all the vogue. People could not hear enough of foreign places and cultures, and the popularity of his guides meant he was in demand as a speaker. When he was invited to present a lecture on Zanzibar to the Church Missionary Society, he determined to enlighten his audience on the weather conditions they could expect to meet, illustrated by diagrams and tables. It was in preparing these visual aids that he came up against the enormous lack of facts from which to work. If they existed at all, they were scattered, undigested and uncoordinated. It seemed obvious to Galton that, if scientists were ever to make sense of the various climates of the world, then they would need access to the right quantities of the right kind of accurate, reliable data. His intellectual appetite was immediately whetted. He had discovered a gaping hole in man's knowledge that begged to be plugged. For an explorer of the land, the challenge of exploring the atmosphere was irresistible and meteorology became his new crusade.

Even without the baggage of family connection and religious belief, as soon as Galton entered the same arena as FitzRoy, the two were set to clash, battling from either side of a deep philosophical divide. Galton, like his cousin Darwin, had inherited a fair slice of Grandfather Erasmus's imagination. He targeted the everyday world, searching and enquiring until he spotted a need that demanded a solution. Then he set about providing it. From invention to exploration, from the art of bivouac to the science of eugenics, the subjects he tackled were many and diverse. His methods, though, were always the same. Meticulous and logical, as might be expected from the man who would later become a founding father of statistical science, he would narrow the problem to a series of issues, then work towards the solution through a set of predefined steps, each rigorously tested and proved before progressing to the next.

FitzRoy was the complete opposite. He would ricochet from one enthusiasm to the next, dropping one as suddenly as he had adopted it when a change of tack promised to take him more quickly to his destination. All his experience as a sea captain had tuned his mind to work in this way. In the face of an emergency there was no time for considered evaluation of the situation; lives depended on instant and instinctive decisions. When an action felt right, it was right. There was no time to justify it, and in the environment of strict naval discipline, there was no cause to justify it, either.

Galton's circular was merely the first expeditionary scouting in what would be a long campaign. Although there was nothing obviously confrontational in it, FitzRoy did well to regard it as an antagonistic move, for that was precisely the vein in which it was issued. Galton was driven by sheer frustration at the lack of material being released by FitzRoy and the rest of the world's meteorological establishment.

A scientific study of the weather on a worthy scale seems to me an impossibility at the present time from want of accessible data. We need meteorological representations of large areas, as facts to reason on, as urgently as experimental data are required by students of physical philosophy.

The circular had arrived at exactly the time FitzRoy was planning to go public with his daily weather forecasts, in the summer of 1861, and, while he might inwardly shudder at the implications of it, he was much too preoccupied with steering himself and his office into the vanguard of meteorological science to pay much outward regard to Galton's project. All

Galton was attempting to do was to gather together a series of historical simultaneous observations, the selfsame exercise FitzRoy had initiated three years before, and now considered defunct, overtaken by far superior telegraphic data, which would provide ample fodder for historical analysis later. In fact, as Galton was busying himself with his quest for simultaneous observations, FitzRoy was scaling down his own operation, which was soon to cease altogether. And, when his forecasts started, his defences would be raised against the brickbats that followed rather than in fending off an illusory future threat.

Indeed, along with the forecasts came the advent of that popular public pastime, the game of grumbling about the forecast and blaming the weatherman, which we still indulge in today. FitzRoy, being the very first weatherman, was the first to feel the brunt of it. As the editor of *The Times* put it: 'The public have not failed to notice, with interest, and, as we much fear, with some wicked amusement, that we now undertake to prophesy the Weather for the two days next to come.' Failures FitzRoy was bound to have, but in the first week of April 1862 these happened to be very conspicuous, which the same *Times* leader appeared to take a malicious delight in pointing out:

> During the last week, Nature seems to have taken special pleasure in confounding the conjectures of science; and while we, resting upon grave authority, have been made to promise fine Weather, the heavens have bestowed upon us a week of fog and deluge.

The editor was quick to lay the blame elsewhere: 'we are only the mere "medium" of prophecy' but very fairly allowed that 'the failures do not in any degree detract from the importance of the labours in which [FitzRoy] is engaged'. In fact, the editor was very kind to FitzRoy, allowing him to explain his methods to the reader.

> The facts now weighed and measured mentally, in what may be correctly called 'Forecasting' Weather, are the direction and force of each air-current, or wind, reported telegraphically to the central station in London from many distant stations, their respective tension and temperature, moisture or dryness, and their changes since former recent observations.

There is no doubt that some of FitzRoy's phraseology could benefit from a makeover by the Plain English Campaign, and whether his rambling explanation did anything for the reader's confidence, other than overwhelm him with science, is doubtful. *The Times* editor, though, thought he knew what FitzRoy intended, for he went on to attribute to FitzRoy an amazing ambition – that he would soon be able to produce his forecasts, not by mental estimate, but by mathematical formulae. And what is more, 'the generation now newly born may learn at school to work out the Weather of tomorrow as surely as generations, past and present, have calculated eclipses'.

We may smile at such innocent faith in the power of FitzRoy, and it is doubtful that FitzRoy himself believed the procedure could be quantified so quickly to such a precise degree. Yet, in reading that, he must have felt a very heavy expectation placed upon him to perform better, to achieve ever greater accuracy. *The Times* leader concluded:

> there is no reason why the Admiral should be discouraged by anything that has yet happened. Science is acquiring new arms every day, and they who would follow her faithfully need not utterly despair of wresting even such a secret as this from Nature.

And perhaps this gave FitzRoy heart, for soon afterwards he was indeed attempting to help Science acquire one of those brand new arms – and this one he felt would surely wrest that elusive secret from Nature.

Indeed, the summer of 1862 found FitzRoy hatching yet another theory that would revolutionise the world's thinking on the behaviour of the atmosphere – or, at least, he was convinced that it would. His previous forays into science with his theories of atmospheric circulation had not turned academic heads in the way he had hoped. But this time he had no doubts. When news of this was released, Francis Galton could do as he wished; Robert FitzRoy's name would outshine him. Those who sneered at his forecasts would be forced to backtrack once this idea was brought into his calculations and the rate of accuracy shot up as a consequence. It was a brand new idea – well almost – although it had its roots way back in history.

People had long ago thought that perhaps the moon had something to do with the mysteries of the weather, and in an attempt to fathom exactly what, they had tried to find a connection between the weather and the moon's phases. Did a new moon herald a change in the weather? Was the weather more benign under the influence of a waxing moon or a waning moon? People observed, drew up tables – even Luke Howard had investigated the lunar effect – and, by contorting the results to fit the argument, they had made vague deductions. Yet predictions based on them were no more likely to be right than those founded on country lore, or astrology, or even pure chance. But the idea would not go away. In 1846 Colonel Sabine had tried a different tack, postulating that maybe the moon exerted an attraction on the atmosphere just as it did on the sea, producing what he termed 'Lunar Atmospheric Tides', which might create currents in the air in the same way that oceanic tides do in the sea. However, since nothing was forthcoming in the way of proof, there were no further developments. Until, that is, nearly sixteen years later, in 1862, when a scientist by the name of Lieutenant Langham Rokeby boarded a ship and sailed off on the long voyage to Ascension Island, where he was to make observations in order to prove or disprove, once and for all, whether there were tides in the atmosphere caused by the pull of the moon. As soon as his early observations began to flow back to England, Sabine was regaling FitzRoy with speculations on what they might eventually prove.

Sabine's idea clicked with a notion FitzRoy had never quite got out of his head. Ever since Henry Sorby's innocent letter on the action of waves had prompted FitzRoy's early conjectures on atmospheric waves, he had clung to the belief that there must be some similarity of behaviour between the mass of water in the earth's oceans and the mass of air that surrounded it. He had toyed with but soon dropped the theory that the atmosphere moved in waves – though many, including Sir John Herschel, still clung tenaciously to it. Sabine's hypothesis was different. Here, the concept was of a tide, rather than waves, which would cause currents in the atmosphere as it was pulled over the earth's surface by the tug of the moon's gravity. But the theory stumbled on the hurdle of proof. Lunar tides, if they existed, would follow the lunar cycle and so, like every other lunar theory, depended on identifying a similar cyclic pattern in the weather. History had shown that this was almost impossible to find.

FitzRoy, though, by going back to the sea and its tides, thought he had found the answer. Sea tides are predominantly under lunar influence. However, the sun, too, produces a solar tidal effect that causes the strict lunar pattern of sea tides to vary. What if the sun exerts a similar sort of pull on the atmosphere? If this were so, the lunar pattern everyone sought to find would be knocked off keel by the sun's influence. The challenge was to re-examine weather data, analysing it not just with the moon's phases, but the sun's seasons, too. The idea so excited FitzRoy that he threw all his energies into working on perfecting this new theory, which he termed his 'lunisolar' theory.

Meanwhile Francis Galton was hard at work, sifting the results of his survey for presentation in his proposed new book *Meteorographica*. With the book, Galton was introducing a new discipline. As might be expected from a geographer, the book featured maps, and the term 'meteorography' means, according to him, 'a branch of Meteorology which consists of the tabulation of observations, then delineating the observations in a pictorial form'. *Meteorographica* was a work on a grand scale, containing 600 such pictorial maps of the weather over Europe during the month of December 1861. Much more than anything that FitzRoy produced, these maps resembled what we would today recognise as weather charts. Galton introduced to his maps the concept of lines, like contour lines, to link together areas with similar measurements of pressure and temperature – the 'isobars' and 'isotherms' that we are accustomed to seeing on charts today. Unfortunately, his circular had not reaped in the amount of raw data he had hoped for. Observers, already sending their readings here, there and everywhere, were not prepared to address them to yet another destination, especially as Galton imposed so many conditions on them – the postage must be prepaid, the readings 'reduced' (i.e. to sea-level values), temperatures converted to Fahrenheit, and so on and so forth. Individual, small-scale observers on the whole were not interested, and the bulk of his data came from government-backed observatories: from Quetelet in Belgium, Buys Ballot in Holland, Kreil in Austria and the Prussian master of the storm, Heinrich Dové.

Galton was understandably bitter about this lack of response, and in bemoaning it, he descended into outright censure of Robert FitzRoy. 'Meteorologists are strangely behind in the practice of combining the materials they possess,' he wrote, and, damning FitzRoy with faint praise, he added: 'A debt of gratitude is due to Admiral FitzRoy for his methodical daily reports, but I believe them to be insufficiently numerous, extended or frequent.' Yet it was not only FitzRoy's department that was at fault, for the same thing was happening, or rather not happening, all over the world. There were, he pointed out, over 300 trained observers transmitting data daily to meteorological societies or government institutions, yet there was no means for the independent meteorological researcher to gain access to these returns. Any 'real student of meteorology' needed to engage in the laborious and expensive work of personally canvassing for observations, and then combining them, before he could begin any proper scientific analysis.

Such paucity of data could have been disastrous – a few isolated jottings, which in a quiet weather month would have produced a jumble of lines going nowhere, proving nothing. But Galton had the inconceivable good luck of having plumped for a month of fast-moving weather studded with contrasts. Cyclone followed cyclone in a swift passage from west to east, with areas of high pressure sandwiched between. Winds blew hard, then dropped to nothing, and picked up again as suddenly, weaving this way and that in varied directions. Therefore, his maps, though based on such small amounts of data, began to reveal some amazing findings. As he plotted his wind directions on top of his isobars, he spotted a phenomenon that had not been noticed before: just as winds piled into an area of low pressure in a spiralling, anticlockwise direction in the northern hemisphere, they spilled out from an area of high pressure, falling away from the centre in a clockwise direction. If the first was a cyclone, then what he had discovered was its opposite, an 'anticyclone' – so that is what he called it, and that is what it has been known as ever since.

His maps proved, too, the enormous scale of atmospheric movement. Not once did he find the centre of a 'low' and the centre of a 'high' at the same time on the same map. Since even Europe itself was not big enough to contain them, then figures from England alone, no matter how many, how regular and how accurate, were never going to prove anything. What was needed was cooperation on a worldwide scale – an operation of Maury-like proportions. Tenacious, just like FitzRoy when he had a burning objective to achieve, he wrote a pamphlet, entitled it *Meteorological Instructions for the use of Observers Resident Abroad* and, under the auspices of the Meteorological Society, of which he was now a member, dispatched it off to as many folk in as many countries as he could reach. Along with it went a set of meteorological instruments: a maximum and minimum thermometer, a normal thermometer, a rain gauge – but no barometer. FitzRoy must have been aghast. Temperature without air pressure, wind without air pressure, precipitation and cloud cover, also without air pressure – what was the man thinking of? They were all totally meaningless and of no value in helping to understand the movement of the atmosphere and the behaviour of climate.

Meteorographica was undoubtedly a magnificent effort, though no matter how many maps Galton drew, no matter how detailed his contours and wind lines, there was no getting away from the fact that the entire edifice was built on incomplete data – a failure that left his credibility and his conclusions open to criticism. And criticism he got in plenty – none of it more scathing than FitzRoy's.

In December 1862, before publication of the book, Galton presented his findings in a talk to the Royal Society. If FitzRoy was sitting in the audience that day – or even if he was not, but heard of it later – he must have seethed with indignation at what he was hearing. 'Anticyclone' indeed! And as for his isobars, they might as well have been drawn from the imagination. You could not simply plot a line between two sets of observations, hundreds of miles apart, that appear to share the same reading, and assume that at every point along that line the barometric pressure will be exactly the same. He was to write later, to fellow meteorologist J.D. Forbes, who had dared to praise Galton's results:

His views about *anti*-cyclones seem to myself – among many – quite unsupported by facts or any plausible theory. Isobarometric curves – to be really useful, ought to depend on many more stations of observation than he, or even M. Marie Davy have used: I think you will allow when duly considering effects of high land – depressions – snowy ranges of mountains, and local precipitations: besides the direction of winds. Pray consider how little is known of the elevations (above a *sea* or other normal *level* –) of almost all continental *stations inland*.

If Galton had kept entirely to theory, then FitzRoy might have consoled himself with a fit of self-righteous railing against his views. But there was worse, much worse, when Galton turned to fire directly at FitzRoy a personal attack aimed against him and his department. He declared that, on his own behalf, he had obtained, digested and published as many discrete observations for one single month as the Board of Trade had managed to issue in its weather reports for an entire year. Perhaps, he dared to suggest, instead of an annual regurgitation of already published telegraphed weather reports, the Board of Trade should print an extended list of observations 'taken at leisure'. In other words, rather than giving us what we have already got, FitzRoy should stop sitting on the backlog of materials he has stowed away and produce them for all to use. Not content with criticising, he ploughed on, instructing FitzRoy in how he should do his job. Produce the figures, not in printed columns, he said, but in a more compact, partially digested form – perhaps even in a series of small maps, exactly like those used in *Meteorographica*, but extended to cover a period of two to three years. They would afford the means, he helpfully pointed out, of 'testing the extant theory of "forecasts" with a rigour impossible at the present time'.

So Galton was casting snide aspersions at FitzRoy's weather forecasts – and not only that, he had the audacity to slate his storm-warning predictions, too. Galton claimed, on the strength of his extremely limited sample of data, to have proved that storm warnings of adequate value could be telegraphed to a whole line of coast from a central office, whenever a gale is *known* to be blowing. FitzRoy had a right to feel affronted. Forecasts were one thing – he was becoming accustomed to defending them – but his storm warnings were accepted by most of the population as being a valuable service. Even *The Times* in its April leader had praised their worth. So how could Galton pit his ridiculously puny sample against FitzRoy's breadth of knowledge? With all his years of observation, with the possession of the wealth of data Galton himself had accused him of hoarding, with all his practical experience at sea, and with his near two years of successfully *predicting* storms, who could know better than FitzRoy?

It would take a man of supreme self-confidence to turn the other cheek to such an onslaught, and FitzRoy, for all his arrogance and bluster, was not such a man. Just when he was beginning to win the respect he craved as a meteorological scientist, just when he had gained public approbation and thanks, his reputation was being torn apart by Francis Galton. People would take note of Galton – he was an eminent scientist in his own right, and a Darwin to boot. He had FitzRoy on the defensive, and FitzRoy would defend his methods and his reputation with all the means at his disposal.

As fate would have it, a more immediate put-down for Galton had already fallen into his hands. Some days earlier, just before Galton had announced his *Meteorographica* findings to the Royal Society – and no doubt prompted by the thoughts of FitzRoy that preparing his paper had given him – Francis Galton had contacted Robert FitzRoy at the Meteorological Department, asking him what had been the outcome of the Admiralty's investigation of his heliostat. Chasing it had produced only the return of the diagrams that, come January, were sitting on FitzRoy's desk. If anyone had actually tried it, they had not sent back a report. FitzRoy did not trouble to ask. Instead, he had the plans packed up and it was no doubt with a wry smile that he penned the insultingly brief note to go with them: 'The instrument does not seem to be thought practical.'

Although that small action no doubt gave him immense personal satisfaction, he needed a far wider arena in which to display his superior standing. Contorted with fear of suffering a public pounding at the hands of Galton when *Meteorographica* was published, he determined that his sparkling new 'lunisolar' theory should hit the bookshops way ahead of it. For months he had been collecting and collating all his previous works on the weather, tracing the historical basis of current meteorological thinking, adding new developments, and writing up the sum of his experience, all for publication under the title *The Weather Book*. It was to be a work aimed not at meteorological scientists, but at a broader audience of everyone who took an educated interest in the weather. It might not be the ideal vehicle in which first to air his new, highly technical, theory, but it would be the quickest, since the manuscript was nearing completion and a publication date agreed for the spring of 1863. Ready or not, proof or not, it was imperative that an additional chapter should be slotted into *The Weather Book*. He spent the rest of that December writing furiously and managed to finish the manuscript just before Christmas.

All that remained, for his own peace of mind, was to gain the stamp of approval for the book as a whole, and for his lunisolar theory in particular, from a scientist he admired. Proud of his efforts, he parcelled up a proof copy of the book and posted it on Christmas Eve, an unusual Christmas present for its lucky recipient.

Burden of Proof

I cannot dismiss the idea that there is something more than nonsense in the new part – till you put down your foot upon it . . . and entirely condemn the notions.

The man blessed by FitzRoy with this power of veto over his lunisolar theory was Sir John Herschel. He was a strange choice. The two were often at loggerheads. Their so-called cooperation on the lighthouse scheme had been fraught with argument, and the atmospheric-wave controversy was still involving the two in sharp exchanges. But, despite all this, Sir John was one of Britain's most highly regarded scientists and FitzRoy, like most others, held him in great awe. And then again, it had been Herschel who, back in 1836 when the *Beagle* called in at the Cape, had urged FitzRoy to watch the weather in conjunction with the phases of the moon and to note all the coincidences. Sir John was a staunch lunarist, a study he had inherited from his father and had keenly pursued while at the Cape. Based on his enthusiasm then, Sir John could now be expected to look favourably on FitzRoy's innovative new slant on the old theme. FitzRoy certainly expected him to do so.

It was one thing to spout his emergent theory onto the pages of a book, but proof was quite another – and when the book was published, the clamour would be for solid data to back up his claims. Despite the fact that he had no authority or mandate to conduct an official investigation, early in January he reorganised the work of the department in order to free Babington to start the urgent task of coordinating and plotting the observations needed. Babington was gradually becoming FitzRoy's right-hand man. In his work as a clerk, he evidently displayed sympathy with FitzRoy's aims and an understanding of the various flights of theoretical fancy that constantly tumbled from FitzRoy's mind.

As 1863 began, FitzRoy was entering the most stressful period of his meteorological life so far. February marked the first anniversary of his storm warnings, and the end of the experimental phase. It was time for a decision to be taken about their future – would they be adopted officially, or would they be withdrawn? Added to that was the vexed question of extra finance for the telegraph system. That particular 'experiment' had technically finished the previous autumn, yet a decision was still awaited on whether to continue funding. Meanwhile, the system kept running, storm warnings were issued and weather forecasts made. And it all cost money. How long before the Chief Accountant stepped in and kicked up a fuss? To cap it all, he had not had a reply from Sir John Herschel, and publication of *The Weather Book* was imminent. Was he to sanction the go-ahead before confirmation of Sir John's blessing, or would he publish without the controversial new chapter, and see his name tarnished by Francis Galton's *Meteorographica* before he had made his mark? Since this last decision was his to make, he made it. He authorised publication of his *Weather Book* complete with its lunisolar chapter. And he remained hopeful of Sir John's positive endorsement.

At least the vibes coming from France were positive. He had been contacted by a French naval officer, Captain Mouchez of the Ministry of Marine, the French equivalent of the Board of Trade's Marine Department. FitzRoy, in fact, already had a close working relationship with the Ministry of Marine. This was the ministry that had been charged with the administration of the observations required as a result of the Brussels Conference. A Captain Moulac, Commander of the Naval Station at Honfleur, had visited FitzRoy on behalf of the ministry the previous summer, in order to discuss arrangements for the transmission of telegraph weather reports direct to the ministry, in addition to those he was already sending to Le Verrier at the observatory. Now the French Navy had grown impatient with waiting for a storm-warning system to emerge from Le Verrier, and Mouchez had been authorised to come to a separate arrangement with FitzRoy for the transmission of his storm warnings. Despite his pretensions to scientific status, FitzRoy was always happier dealing with navy people, especially those who regarded him with respect, and in these French officers he had found two avid supporters. Already, in January, less than two months after the arrangement had been in place, Captain Moulac was enthusiastically reporting that FitzRoy had transmitted advance warnings of three gales that had hit France. Since the Board of Trade had set such store by French opinion when it had come from Le Verrier, then perhaps they would do the same with this.

Unfortunately, they did not. The President of the Board of Trade, Milner Gibson, announced his decision on 12 February. The good news was that the experiment in the telegraphic communication of meteorological data had proved itself, thereby justifying the increase in expenditure. The bad news was that, with FitzRoy's storm warnings, Milner Gibson had opted for the usual 'get-out' route, recommending that, since the usefulness of the warnings was uncertain, the Royal Society should be consulted and that a programme should be instigated to test accuracy by comparison with actual weather.

If FitzRoy was aghast at all this, he had every right to be. Why ask the gentlemen scientists of the Royal Society if they had found the warnings useful? How many of the members earned their livelihoods risking their lives at sea? The people to ask were the small-scale sailors and the fishermen – the very people FitzRoy had already surveyed, and who had provided their favourable answers to the Board of Trade. As if that were not bad enough, he must have been truly incensed when he discovered the treachery taking place behind his back, perpetrated by the duplicitous Thomas Farrer. Just as he had, three years earlier, taken it on himself to tag Le Verrier's warning about storm predictions onto FitzRoy's proposals to the Board of Trade President, he now 'enhanced' the President's referral to the Royal Society with an addendum strictly his own. He asked it to consider whether the money presently being laid out on storm warnings would be better spent in running the department according to the Royal Society's own original brief.

As for the proposed programme of comparisons of warnings against actual weather, FitzRoy and Babington had been carrying out that same exercise since the previous summer. Not unexpectedly, FitzRoy's own results showed a marked degree of accuracy – but what was not accurate was the method by which he had arrived at the results. He simply used the regular daily transmissions of weather data, augmented by newspaper reports, which he compared with his record of warnings issued. Now, following Milner Gibson's edict, the Wreck Department (another department of the Board of Trade, quite separate from the Meteorological Office) would undertake independent monitoring, and a task force of special observers was to be employed to watch for actual storm conditions, or the lack of them, in areas where warnings were in force. Not surprisingly, FitzRoy was not happy with this arrangement, since it was totally outside his control.

Comparing actual with predicted sounds straightforward. Say, for example, that a storm cone was hoisted at Newcastle on Monday, and on Thursday it was removed. The question is, did a gale actually occur at Newcastle on any of the three days – Monday, Tuesday or Wednesday? If yes, the warning was accurate, if no, it was inaccurate. Unfortunately, measurement was not nearly so easy. For instance, take the cone itself. It warned not just of an approaching gale, but also of the expected direction from which it would come. If the storm was predicted to arrive from the north and it came from the south, could the warning still be counted as accurate? Then there was the problem of how many signals constituted one warning. When FitzRoy sent a storm warning to Newcastle, he would usually send a similar warning to stations in the same area, because, since he could not guarantee to pinpoint exactly where the gale would hit, he liked to spread the warning to be on the safe side. Now what degree of accuracy would be recorded if just one of those stations returned winds of gale force 8, and the others recorded winds of only force 7? According to FitzRoy, he would have achieved what he set out to do, and he would claim total accuracy. On the other hand, it could be said that, of the several warnings transmitted, only one warning was correct and all the warnings sent to the other nearby stations were incorrect – that is, four or five 'wrong' warnings to the one 'right' one. Less than 20 per cent accuracy, his detractors could claim, with statistical, though not necessarily logical, justification. Now factor in the human element, in that wind speeds were measured not by instrument but by eye, add the fact that observers were stationed on land, where the wind was less violent than conditions at sea, and the position becomes even more uncertain. It could be that some of those force 7s should really have been returned as 8s – or, on the contrary, that the one 8 was not an 8 at all, but only a 7. Like all statistics, without a strict predetermined yardstick for measurement, the figures were open to the interpretation that best suited the user's case. And that worried FitzRoy, as well it might.

Undoubtedly the seamen, at whom the warnings were aimed, supported FitzRoy and found him to be accurate more often than not. But what weight would their voices carry against the more eloquent and influential opinions of his detractors? According to FitzRoy, his critics fell into four categories. The first he quoted as being 'certain persons' who had been opposed to the scheme in theory, right at the beginning, and who were reluctant to change their minds. Among these might be numbered Francis Galton, who objected on scientific grounds, along with bureaucrats like Thomas Farrer and others, for whom, as laymen, a system of reporting only the definite existence of a storm rather than the 'maybe' of prediction seemed to make much more sense. Secondly, there was the 'numerous body' of the uninformed public, who did not understand the need for or the rationale of warnings and forecasts, and who simply liked to poke fun, calling the whole thing a 'burlesque' and valuing it no more than fairground fortune telling. For his fourth category he quotes 'those pecuniarily interested individuals, heedless of the precarious occupation of the Coasters and fishermen'. By these he meant the fat-cat shipowners, sitting safely sipping brandy in their gentlemen's clubs, bemoaning the dent in their pockets owing to a delay in landing a catch of fish, and all because the sailors preferred not to risk their lives in the face of a FitzRoy Cone. His third category was the very small yet extremely vociferous group led by James Glaisher.

Glaisher was the man who had joined the Balloon Committee of the British Association at the same time as FitzRoy. FitzRoy had dropped ballooning almost as quickly as he had taken it up. Once he had gone public with his theory of atmospheric circulation, the burning need for corroborative evidence faded, and his interest flitted onto his next enthusiasm. Not so James Glaisher. He was as passionate about meteorology as was FitzRoy, but, whereas FitzRoy was intent on harnessing the science to save other people's lives, Glaisher was prepared to risk his own life in pursuit of the science. To be fair, both Lord Wrottesley and Glaisher had tried to keep FitzRoy involved, inviting him to various balloon launches and lectures, but he had declined them all. Yet while FitzRoy was content to burrow under his figures in his Whitehall den, the intrepid Glaisher itched to get airborne. FitzRoy's rejection would not have pleased him, since there was a certain prestige in having the FitzRoy name associated with the venture.

Glaisher's ambition was to fly higher than anyone had flown before. The difficulty was in finding a balloon fit for the task. Balloons are expensive now, but in the 1860s, when sheets of delicate material required miles of seams hand sewn by teams of skilled tailors, the cost was horrendous. Not many were around, and those that did exist were either too old and worn or simply not big enough. Veteran balloonist Henry Coxwell, by then in his mid-seventies, volunteered to have a specially designed balloon made and to pilot it himself. Fortunately, the British Association agreed to continue its backing of balloon meteorology and came up with the necessary finance. But in return for its money, it expected spectacular results and Glaisher was under pressure to provide them.

Sponsored by Lord Wrottesley, the Wolverhampton Gas Company again provided the gas – a whole gasometer full of light gas, specially saved for the purpose. Together with Coxwell, Glaisher made his first ascent on 17 July 1862, rising to the incredible height of 26,000ft, or nearly 5 miles above the earth. Intoxicated with the experience, Glaisher hankered after even greater height, and he persuaded Coxwell to take him up again. On 5 September the pair clambered into their basket, along with a barometer, thermometers, hygrometer, a camera, six pigeons and a bottle of brandy. The balloon, called *Mars*, rose up over the Staffordshire countryside and disappeared into thick cloud. After another 1,000ft it burst out above the cloud and continued spiralling upwards at a heady rate. Forty-five minutes after leaving the surface, they had reached a height of 5 miles, where the air temperature had dropped to −5°F (−20°C), and their problems began.

The first intimation of trouble was when Glaisher found he could not focus his eyes on the fine column of mercury in the tube of his thermometer. He looked to Coxwell for help, but to his dismay the man was clambering out of the soaring basket, clinging precariously to the ropes with nothing but a cloud between himself and certain death 5 miles below. Glaisher passed out. Coxwell meanwhile was struggling to free the valve line, which had twisted and was caught up. He managed it, and slid back into the basket, but his hands were so frozen he could not use them. The line was free but the balloon was still accelerating upwards and both men were within seconds of certain death. Somehow, just before he, too, passed out, Coxwell managed to catch the valve line with his teeth and jerked hard. The balloon ceased its upward rush and descended into warmer air. James Glaisher described the next scene of the drama in his 1871 book *Travels in the Air*:

Whilst powerless I heard the words 'temperature' and 'observation', and I knew Mr Coxwell was in the car, speaking to and endeavouring to rouse me, – therefore consciousness and hearing had returned. I then heard him speak more emphatically, but I could not see, speak or move. I heard him again say, 'Do try; now do'. Then the instruments became dimly visible, then Mr Coxwell, and very shortly I saw clearly. Next I rose in my seat and looked around, as though waking from sleep, though not refreshed, and said to Mr Coxwell, 'I have been insensible.' He said, 'You have; and I too, very nearly.' I then drew up my legs which had been extended, and took a pencil in my hand to begin observations. Mr Coxwell told me that he had lost the use of his hands, which were black, and I poured brandy over them.

I resumed my observations at 2hr 7m [seven minutes past two]. It is probably that three or four minutes passed from the time of my hearing 'temperature' and 'observation', until I began to observe.

Guessing what the rate of ascent and descent must have been while he was unconscious, Glaisher claimed they had climbed to a height of 7 miles. This is unlikely, but even at his highest recorded altitude of 29,000ft, the ascent was a world record, which they would retain until the age of powered flight.

Another risky scheme that Glaisher launched himself into was also a flight into the unknown. The year 1862 saw a piece of landmark legislation come onto the statute books – the Limited Liability Act, which created the concept of the limited company. One of the ventures to emerge into the first rush of such companies was the Daily Weather Map Company

(Limited), with James Glaisher as its founder. The idea was to set up in opposition to FitzRoy, to do exactly as he did in accumulating a mass of telegraphed weather data and publish the results, but in Glaisher's case for profit. In 1863 the company issued a prospectus, inviting people to buy shares in the venture.

> Much has been accomplished under the supervision of the Board of Trade, and by the aid of an annual Parliamentary vote, with highly useful results. In the *Daily Weather Map* the process will be carried out far more fully and extensively, the observations collected at a large number of stations in different parts of the British Isles being telegraphed to London, and reproduced for publication within two hours, not only in tables which are intelligible to only a few readers, but also delineated on a Map or Chart, so as to appeal to the eye, and present a sort of profile outline of the wind and weather as existing at 9 o'clock every morning in every part of the United Kingdom to which the means of telegraphic communication are now provided.

The scheme would be bigger, faster, and more cost effective than FitzRoy's since, instead of gobbling up public money, Glaisher's scheme would produce money for the public – the investing public, that is. Ambitious it certainly was. And clever. Recognising that the punters would not want to pay out extra for a daily weather map in addition to a daily newspaper, he decided to include news pages with his weather map and, for revenue, to offer advertising space.

Unlike the balloon, this venture failed to get off the ground. Despite Glaisher's own investment in time and money – he had already planned the operation, printed a prospectus and designed and started the manufacture of instruments for the observers – it would appear that investors were not forthcoming. Glaisher had calculated his returns on a daily circulation of at least 3,000 copies, aimed at scientists, mariners, merchants, agriculturists and travellers. But where were all these people? Unless they were in central London – and most of them were not – by the time they received the maps the information in them would be out of date. FitzRoy's forecasts, on the other hand, covered a forty-eight-hour period, and were therefore still of value whenever the newspaper managed to reach distant areas. Glaisher claimed his charts would be easier to understand than FitzRoy's tables. They would need to be. As a true scientist, he refused to provide anything as chancy as a weather prediction. He would merely provide the data and leave it to the individual reader to 'anticipate coming changes, to trace the variations of the wind, and foretell the approach of rain, storms or sunshine'. Hit-and-miss as FitzRoy's forecast tended to be, it was probably more accurate than that which the untrained sailor or gardener could manage on his own from a mass of figures, albeit splurged across a map rather than in printed columns.

When the venture collapsed, Glaisher and his fellow investors blamed FitzRoy and turned against him, criticising his operation publicly in letters to the newspapers. If he was to blame, it could only be for doing his job so well as to make the commercial venture non-viable.

No matter how hard FitzRoy tried to justify himself and to improve his service, the carping critics shouted louder, and the louder they shouted, the more his masters were likely to listen – and all this at a time when the future financial funding for his very employment and livelihood depended on proof of the usefulness and accuracy of storm warnings. Fortunately, *The Weather Book* was promising to become a bestseller, and, although this could do nothing but good for his popular image, it was to stand as a fully fledged meteorological scientist that he craved, and for that he clung to the lifeline of the hurriedly added chapter on his lunisolar theory. He realised that the ordinary reader might not understand it, but he hoped the leading meteorologists would do more than comprehend. He hoped they would applaud. With nail-biting eagerness he awaited the long-delayed verdict from Sir John Herschel. On the morning of 14 March a letter written in Sir John's hand was delivered. His fingers must have shivered with apprehension as he tore desperately to break the seal and unfold the sheets. The first sentence he read was: 'I am sorry to be obliged to say that I must demur to every one of your conclusions in the Gravitating Lunisolar influence on the atmosphere.'

Herschel's rejection crashed into the shaky construction of FitzRoy's theory with the force of a demolition ball, shattering its credibility and creating a dent in his hitherto indestructible self-belief. He fought back, telling Herschel he now had two months of data to support his theories, but the great man was unimpressed. Sarcastically advising FitzRoy to consult a better mathematician, he firmly shut the door to any further appeal on the subject.

Just as FitzRoy was reeling from the viciousness of Herschel's attack, Francis Galton's *Meteorographica* was finally published. This appealingly presented book with its sheets of convincing maps caught the public imagination, while its preface contained all the gall of Galton's bitterness towards him. And still hanging in abeyance was the question of withdrawal of funding for his storm warnings. The Royal Society had still not produced its recommendations, and, just as he had foreseen, the Board of Trade Accountant, H.R. Williams, was breathing down his neck, complaining that expenditure was over budget to the tune of £1,500. The entire future of his system of forecasting depended on House of Commons approval. His health was suffering, the young men on his staff were suffering and all because no one would seek to make a decision.

Nervous stress sent him into overdrive. He strove harder and harder to improve the accuracy of his forecasts, he pushed Babington into gathering more and more observations on lunar gravitational effects, increasing the scope to include records of electrical interference on telegraph wires, and he himself started work on a paper to present to the Royal Society to explain and show proof of his theories. He had made an impulsive promise to drop his own belief should Herschel put his foot on it. Well, Herschel had emphatically slammed both feet down, but FitzRoy could no more afford to 'drop it' than he could afford to stop breathing. With opposition building rapidly, he was left with the only defence open to him – he had to prove through his own investigative efforts that what he had discovered in his lunisolar theory was indeed the major breakthrough the meteorological world was waiting for.

However, while scientists rejected him, the ordinary people revered him. Indeed, his name was rapidly becoming the stuff of legend way beyond British shores.

The French Connection

The meteor was a beautiful sight. In a fit of sparks, it soared across the sky, touching the stars with its tail. The message it carried was as clear as if it had been written in the heavens by an unseen hand. On the calm, blue waters off Oporto, fishermen abandoned their bulging nets and made in haste for the shelter of harbour. In Gibraltar, the quiet streets beneath the Rock were suddenly thronged, as good Catholic folk poured from their houses and hustled to the churches, where such was the crush that the great wooden doors were nearly forced from their hinges. The congregations flopped to their communal knees offering up desperate prayers in supplication to God. Deliver us from the storm, they begged.

Their prayers were answered. The skies stayed blue and the weather remained unusually fine for the time of year. The people, placated, went about their normal tasks once more. But the damage was considerable – expensive catches were lost, labour was withdrawn, and people suffered from shock and injury in the press of the crowds. God had averted his wrath, but the anger of the Gibraltarian and Portuguese authorities was turned squarely on Robert FitzRoy in the fastness of his office in far-off London. It was he they blamed. He had forecast, they said, the severe tempest that would strike Gibraltar and the southern coast of Portugal between 8 and 18 December 1863. And had not FitzRoy himself claimed that the disastrous English storm of October 1859 had been forewarned by meteor? This meteor, then, had to be the sure sign of the tempest's coming.

FitzRoy protested. 'To whom these absurd but injurious predictions are traceable I am not certain, however suspicious.' What he was certain of, though, was that the entire fracas was none of his doing and was indicative of the steps to which his opponents would go to malign him. As he was at pains to point out, it was impossible to forecast more than a couple of days in advance, even in the British Isles, an environment that he studied daily. He certainly would never attempt to predict over a ten-day period, as this warning had done. There were groups who would, though – the astro-meteorologists, who still persisted in the ancient belief of weather prediction via astrology, or the lunarists such as S.M. Saxby, who clung to the old ways of linking the weather to the phases of the moon – and it is possible a warning of theirs had somehow been linked to FitzRoy. Whatever its source, the incident proved just how far his renown had spread by the end of 1863 and just how much trust was put in his weather warnings by ordinary seafarers, even at the other side of Europe.

This European fame had been assiduously courted by FitzRoy over the preceding few months. Jilted by Herschel, pilloried by Galton and hounded by his superiors, he was forced to search elsewhere for support and recognition. He might at one time have unburdened himself to Maury, but Maury had abandoned meteorology to embroil himself in the Civil War.

Maury's position in the US Navy in Washington had become untenable when his home state, Virginia, seceded from the Union in April 1861. On 20 April this prolific writer composed the fateful few words that would mark the end of his career as meteorologist and geographer of the seas – his resignation from the US Navy. He handed over his papers and all his work to his deputy, Lieutenant William Whiting, left Washington with reluctance and a heavy heart, and returned home. His resignation was not accepted; he was instead discharged. His preferred successor at the observatory, Lieutenant Whiting, was passed over in favour of Commander James Gillis, the man originally championed by Bache for the position that, seventeen years earlier, Maury had won. It had taken ten years of waiting, but at last Bache could gloat in victory over Maury. And Gillis's first action was to dispense with Maury's life work – his marine observations and charts – no matter that the merchant fleet and even his own Union Navy relied upon them for their safety at sea. Even the North's newspapers, which had once adopted Maury as their darling, were now vitriolic in their attacks. 'His career . . . will exhibit one of the most remarkable and successful careers of unblushing charlatanism known in the world's history,' blurted the Washington *Daily Star* in his wake, accusing him of 'filching' ideas from Henry, Espy and others, and of having 'trickster adroitness enough' to be able to bend American minds to his cause.

Despite lucrative offers from both Russia and France of sanctuary plus funding to pursue his meteorological research, Maury saw it as his duty to support his state in its war effort. He accepted instead an appointment to the Council of Three, set up to advise the Virginian State Governor in the conduct of the war, and took a commission in the Confederate Navy. In the summer of 1862 he mined the James river, but his plans fell into enemy hands. However, an electric torpedo of his own design made a direct hit on the steam corvette *Commodore Jones* as it made an exploratory sailing up the James. Exploding wreckage shot sky-high, blocking the Union Navy's access to Richmond, a city strategically vital to the Confederates. Maury's weapons were spectacularly successful, but without the requisite raw materials, the Confederates had no means of building up a stock, so Maury was dispatched to England to obtain them. Escaping the Federal blockade was easier said than done, but eventually in October 1862 he managed to get himself and a few belongings onto a ship, along with his youngest son, 13-year-old Matthew.

Once in England, he immediately began campaigning for the Confederate cause. With the help of the Revd F.W. Tremlett, who was to become his closest and most trusted friend, he founded an organisation for the Promotion of the Cessation of Hostilities in America and lobbied leading military, clerical and political figures to sign a petition against the Civil War. It was not so easy as he had hoped. Whereas in the world of meteorology his name was praised, in the arena of general politics it amounted to little. He found himself up against powerful opposition in Gladstone and the influential British anti-slavery activists, all speaking in support of the Unionists. Nothing daunted, he ploughed on with his war effort. Along with Captain James D. Bulloch, he scoured the docks and shipyards of England to source battle cruisers for the sadly depleted Confederate Navy, but the ships they acquired either proved unsuitable, sank before they had even left European waters, or fell prey to Union ships prowling the Atlantic. Meteorology was the last thing on his mind. Immersed in his work and worried about his family back home, the privations they were suffering and the dire tidings of the war filtering across the from America, he had little time for it. He certainly had no time for his former associates such as FitzRoy, a fact that prompted FitzRoy to write to Herschel: 'I have scarcely seen Capt. Maury who is now solely a political.'

With the USA effectively cut off by the war and Maury lost to him, FitzRoy had only Europe to turn to, and it was *The Weather Book* that was to open up the way for him to enter the elite of European meteorology. In the summer of 1863 he received an unexpected boost in the form of a request from Marié-Davy of the Imperial Observatory in Paris for permission to translate the book into French. FitzRoy's relationship with Marié-Davy had set off some months earlier on a most unpromising footing. His dealings with the observatory had always been conducted exclusively via Le Verrier, but Le Verrier, it appeared, was frequently absent from his post just when FitzRoy needed to contact him urgently. Frustrated one day, FitzRoy had sought out the man he assumed to be Le Verrier's deputy, E.H. Marié-Davy. Marié-Davy was outraged at the unintended slur. M. Le Verrier does not have a deputy, he had replied curtly. I am the head of the Magnetic Department of the observatory. Unexpectedly, though, Marié-Davy executed an abrupt about turn and began writing to FitzRoy almost as if he were a long-lost friend. The truth was that, suddenly, he needed FitzRoy.

Le Verrier's 'absences' were probably a campaign of non-cooperation, brought about by his envy of FitzRoy. Despite Le Verrier's attempts to warn him off, FitzRoy had netted himself the kudos of being first with a prediction-based service, which had now been operating successfully for well over a year. Le Verrier was convinced that, had he been given the freedom, and the finance, to develop his own storm-warning system, the glory would have been his. Yet the uncomfortable truth was that he had not managed to get any kind of warning service off the ground, not even warnings based on facts of known storms. And now, the Ministry of Marine had negotiated its own arrangement with FitzRoy. While the British forecasts were fulfilling French marine needs – and for free – Le Verrier's own plans for a French storm-warning system stood no chance of gaining funding and approval. As the final rub, Mouchez had allowed FitzRoy to change the time he transmitted his signals to France. An hour or so's difference mattered little to Mouchez, and the new timing suited FitzRoy. It did not, though, please Le Verrier. The signals from England were vital to his reports in the *Bulletin international* and now they did not arrive soon enough for his planned time of publication. His frustration had come to a head in October 1862, when he complained bitterly to the Ministry of Public Instruction that it was inconceivable for the French Navy to rely on intelligence from a foreign power. The Navy agreed. In Janaury 1863 the Ministry of Marine proposed to allow Le Verrier the use of its telegraph and two officers, if he would provide a service in place of the British one.

Le Verrier had at long last been given what he wanted – only to find that perhaps he did not want it after all. Like FitzRoy, he had envisioned for himself the prestige of a great scientific breakthrough. To beat him now, he needed to be different. He had never subscribed to FitzRoy's methods of forecasting based on temperature and pressure changes linked to wind directions. He still clung to the discoveries Liais had made, of the atmospheric waves that seemingly had presaged the Balaclava storm, and he aimed for a system that would identify spikes in lines of equal pressure as the precursor of a storm. However, success in such a chancy enterprise, especially with the meagre funds he had been allocated, was far from guaranteed, and the damage that failure would inflict on his reputation did not bear thinking about. FitzRoy, who, Le Verrier grudgingly had to admit, was making a fair go of the enterprise, still had to submit to having his name dragged into disrepute by sour scientists and disgruntled MPs. Le Verrier was determined the same would not happen to him. He needed a scapegoat, someone else to take the blame when the brickbats began to fly. With Liais now long gone from the observatory, his attention alighted upon the new head of the Magnetic Department, Marié-Davy.

As soon as Marié-Davy had taken up post, about a year earlier, he had begun to show an interest in the observatory's meteorological work, and had even expressed enthusiasm for the long-awaited storm-warning system. With no previous meteorological background, with absolutely no experience as a seaman and without any qualified assistants to help him, he was appointed the man responsible for the running of the new French storm-warning service. No wonder he regretted his offhand treatment of FitzRoy, since the Englishman and his *Weather Book* were his only hope of being able to pull success out of inevitable failure.

FitzRoy immediately granted permission for a French translation of the *Weather Book*. Wallowing in the joy of discovering an eager acolyte, he treated Marié-Davy to a long discourse on his innovative ideological developments, including, naturally, his theories on lunisolar gravitational influences. Apparently, he viewed Marié-Davy as an unworked canvas, ready and open to receive the full panoply of his vision. 'I tell you these ideas freely – because I am sure you will appreciate them in seeking the truth – unfettered by preconceived views.'

FitzRoy had been awaiting a ready audience and, in Marié-Davy, he seemed to have found one. No doubt in order to prompt FitzRoy for more, Marié-Davy responded like for like, and a cascade of ideas came flooding in free-flow back across La Manche. So free, that he confessed to FitzRoy that perhaps, in response to FitzRoy's own unguarded confessional, he, Marié-Davy, had been indiscreet in revealing his own ideas.

FitzRoy, who had once been so cagey with Maury in alluding to ideas he thought the other might gainfully use, had no such reservations. In a letter that he began by saying 'not so full a reply as I could desire', his response to Marié-Davy came tumbling out, sheet after sheet. It amounted to a résumé of the total sum of FitzRoy's meteorological thoughts, theories and beliefs, and was probably exactly what Marié-Davy was hoping for. Enclosed were his charts of the *Royal Charter* storm, without the isobarometric lines, 'as has been attempted by Mr Galton in his *Meteorographica*, and by others', the use of which FitzRoy told Marié-Davy was quite unjustified. Proudly, he revealed his lunisolar theory, although he admitted he had yet to trace the full significance of its effects. FitzRoy sent off this epistle and eagerly awaited the next instalment from his new-found soulmate. He was to be cruelly disappointed.

In August 1863 Marié-Davy began the French storm-warning system. Once he had gleaned from FitzRoy the gist of his theories, which he absorbed into or rejected from his own, his energies were so totally consumed by his work that he had little time for letter writing, no matter how interesting FitzRoy might find it. He observed the autumn storms very closely that year and, despite the vehemence of FitzRoy's opposite opinion, he concurred with Liais and Le Verrier that the advent of a storm can be detected by distortions in the lines, or 'curves' drawn on a chart between places of equal barometric pressure. His service operated, like FitzRoy's, by telegram from a central control. Each day he constructed a forecast for the next twenty-four hours, which was sent to French ports to be posted up in public places and inserted in the local newspapers. Additionally, the forecast was transmitted to the French Ministry of Marine, and also to various destinations abroad. The *Bulletin* was extended to include a chart of Europe, complete with its curves of equal pressure, and reports of current and predicted weather conditions.

In fact, Marié-Davy was drawing primitive versions of the familiar isobars on today's weather maps, the same lines that Galton had constructed on his *Meteorographica* charts, and that FitzRoy held in such disdain. FitzRoy's reasons for criticising Galton were equally valid against Marié-Davy. Today's isobars are drawn using hundreds of separate readings, whereas Marié-Davy could rely on only a few dozen, and those were only over land, not at sea. The distortion caused by the sharp dip in pressure at the approach of a storm was not so easily seen from a few scattered readings as it is from the picture of tightly packed isobars on today's charts. Inevitably, and as Le Verrier had feared, the paucity of data led to inaccuracy.

Since both FitzRoy and Marié-Davy were sending their forecasts abroad, comparisons were unavoidable. Indeed the French Ministry of Marine still expressed a preference for FitzRoy's, claiming they were more accurate, and despite receiving Marié-Davy's homegrown reports, more and more French ports were clamouring for the return of FitzRoy's authoritative warnings. In his Annual Report for 1864, FitzRoy took pride in reproducing a letter from Denmark, from Forchhammer of the Copenhagen Observatory, which revealed that, while FitzRoy had successfully warned of a severe storm that would hit the Baltic on 10 and 11 December 1863, Marié-Davy had forecast calm. In these circumstances, it was little wonder that Marié-Davy lost the inclination to continue a personal correspondence with FitzRoy.

However, FitzRoy was not idle while he waited for Marié-Davy to reply. Some months before, Faraday had forwarded to him a request he had received from an Italian, Matteucci, for information about the meteorological instruments used by the British, and for copies of the proceedings of the Meteorological Office. FitzRoy had issued a standard reply, then thought nothing more of it – until, out of the blue, Matteucci wrote directly to FitzRoy for advice on the manner of drawing up forecasts. The letter landed squarely in the vacant hole left in his exchanges with Marié-Davy. FitzRoy fell on it with glee, scenting another willing convert. Out came his pen, and the words were flowing again, generally following the pattern of his first letter to Marié-Davy. Only this time, significantly, he did not venture any mention of his lunisolar theory, but instead claimed that it was the general equilibrium of atmospheric current, as described by Dové, that was the key to forecasting.

Matteucci's response was not quite what he had hoped. He dismissed FitzRoy's flood of scientific formulation with the bald statement that he preferred practice to theory – and anyway he had never understood Dové. What he wanted from FitzRoy was a set of simple instructions explaining how to organise a series of meteorological stations for forecasting. FitzRoy was deflated, annoyed and affronted. 'I cannot leave all theory out of sight,' he pouted, 'nor do I find M. Dové *now* difficult to understand'. And if Matteucci could not be bothered, then neither could he – it appeared the man just wanted to be served up with the results of FitzRoy's sweat and toil for free, without showing the slightest willingness to put in a little effort for himself. He packed off to Mateucci Robert Scott's English translation of Dové's book – though how an English version would help an Italian where the original German had failed is difficult to see. Quite probably, FitzRoy did not care.

Meanwhile, he had the French to deal with once again. An article written by an Englishman, Dr Phipson, in the magazine *Cosmos*, had implied that M. Le Verrier of France, back in 1860, had attempted to dissuade FitzRoy from persisting with predictive storm warnings. Le Verrier was incensed and denied he had done any such thing. Where is this letter I am supposed to have sent? he demanded of FitzRoy. Patiently, FitzRoy tried to placate him. He could not risk upsetting the French while he still hoped to gain backing from them for his theories, yet he could not let Le Verrier wash his hands of his culpability. He had indeed said exactly what Phipson accused him of, and FitzRoy must still have smarted over the obstacle course of hurdles he had had to surmount at the time, all due to Le Verrier.

At this point, in the midst of fending off Le Verrier, worrying over the continued silence of Marié-Davy and wondering whether the two might be connected, he suddenly found himself on the receiving end of yet another unexpected communication from a Frenchman. M. le Maréchal J.B.P. Vaillant, the Emperor's Minister of War, contacted him, seemingly about storms and forecasts in connection with the armed services. Although FitzRoy was not to know it, Vaillant's approach was as a result of a verbal tug of war currently waging between the Imperial minister and Le Verrier.

On 7 December, some two months earlier, Le Verrier had presented to the French Academy of Science the first report on his observatory's meteorological work, written in the name of Marié-Davy. In his presentation, he praised the new storm-warning service, quoting the example of a ferocious storm that had hit Western Europe at the beginning of December, the course of which, he claimed, had been successfully traced and warned of by the observatory. Suspicious of these assertions, Vaillant took away the report and studied it closely. At the next meeting of the academy, before an influential and sceptical audience (forecasting was accorded much the same disdain by French scientists as it was by the British scientific community), Vaillant laid into Le Verrier, almost accusing him of lying over the accuracy of storm prediction. That storm, the one Le Verrier had been so proudly proclaiming, was not evident in the curves of equal barometric pressure plotted by the observatory, said Vaillant, and the weather forecast in the *Bulletin* showed not a hint of the coming storm. This criticism of a service still in its infancy does perhaps seem cruel, but it came as a counterpoise to Le Verrier's pompous self-congratulatory attitude and his blatant overestimation of its successes. Le Verrier was under attack, and in normal circumstances he would have hit back with a stinging retort. But his persecutor had the ear of the Emperor, and Le Verrier's very existence as head of the observatory depended on Imperial patronage. Fear diminished his defence to a polite reference to the full, and not just the published, charts of the day in question, and the telegrams of congratulation he had received from ports where shipping was saved from wreck by his timely warning.

In lambasting Le Verrier for relying on the use of isobars for forecasting, Vaillant was reflecting FitzRoy's dismissal of the method. FitzRoy's influence was evident, too, in the theory put forward by Vaillant to the academy that the origin of a storm came in the clash of polar and equatorial wind currents, rather than atmospheric waves. Perhaps it was for support of these views that Vaillant wrote to FitzRoy a couple of weeks later, canvassing his opinion. Or perhaps he already had it in mind to curb Le Verrier's meteorological activities, and simply needed the ammunition to do so.

FitzRoy accepted Vaillant's apparently innocent enquiry at face value, and took up his pen, grateful for the godsend of another promising outlet for his theories. After an informative discourse on the collection of weather observations and the nature of Atlantic storms, he tagged on, in reply to the specific request for his views on the defence application of forecasting, a rather unguarded remark about naval and military commanders preferring not to hear warnings and alarms. Vaillant reported FitzRoy's thoughts back to Napoleon, unfortunately including this remark. It must have taken on an extra twist in translation, for when Vaillant went on to publish the gist of his conversation with the Emperor in the *Bulletin*, FitzRoy's statement had gained the weight of an official opinion and created a furore. Captain Mouchez at the French Ministry of Marine, FitzRoy's opposite number in Paris and a keen supporter, was appalled at what he was sure was a misrepresentation, and he set off a ping-pong of communications that ricocheted back and forth across the Channel. Have you seen this? he asked FitzRoy, at the same time demanding of Le Verrier how come the 'off-the-cuff' remark attributed to FitzRoy was ever printed in the *Bulletin*. On quite possibly the same day, there arrived in Le Verrier's hands, from FitzRoy, a copy of the disputed letter in which he had told FitzRoy not to engage in storm prediction. It was a very timely arrival. Le Verrier could no longer deny he had indeed written the words he was accused of, and had no wish to be slated publicly as a liar yet again. The safest route was to attempt to make amends. So he apologised to FitzRoy. He explained that he had read the Napoleon passage before publication, saw that it could be embarrassing to FitzRoy and had it struck through before going to print. He was stupefied to hear from Mouchez that it had been included after all. It was an unlikely explanation, and no doubt FitzRoy was not fooled. Still, he did manage to shrug it off, partially because the lull in the interchange of letters with Marié-Davy was now being filled with a comfortingly frank, private correspondence with Mouchez.

There is a sense of desperation in FitzRoy's courting of European, and especially French, approbation for his theories at this juncture in his career. He felt, and with some justification, that the scientific community in Britain, influenced by Galton and Glaisher, regarded him as a 'quack' practitioner – all right for dispensing his storm warnings as a sort of balm to appease the common mariner, but not to be taken seriously as a player in the top team. America, where he might once have found consolation in the approval of a world-renowned figure such as Maury, was now lost to him, engaged as it was in tearing itself apart in the Civil War. Apart from the Dutchman Buys Ballot, and the Prussian Heinrich Dové, neither of whom FitzRoy appeared ever to have had much contact with – he seems to have regarded Dové, especially, as inhabiting a higher plane than himself – the French were now second only to Britain as the leading nation in meteorological science. To be hailed with praise there would give him that sense of worth that he had spent his life chasing.

But just when he felt he was achieving the acceptance he craved, the *Bulletin international* was lining up another slanderous insult. This one was to be doubly hurtful, because it came aided and abetted by the man he had long considered his friend and ally – Matthew Maury. It was totally unfair. As soon as he had read it, FitzRoy went to vent his exasperation on Babington, his assistant.

I have just read the Bulletin for this day. How little the writers, one and all, know of the real atmospheric conditions? Does Maury think that a few *revived* notions – does Le Verrier suppose that a hasty and *unpractical* view – ought to overrule all the daily and nightly work given by you and me to this extensive and complicated subject during many years?

After three years of absence from the meteorological fray, Maury had suddenly burst forth with an opinion. Based squarely on his outdated knowledge – or 'revived notions', as FitzRoy so aptly put it – he ventured to castigate FitzRoy's attempts at forecasting, blaming the 'disadvantage of forecasting in an insular position'. 'Nothing could now be more fallacious,' said FitzRoy. There might have been a forgivable reason back in 1861 to believe that the only way to forecast a storm was by advance warning from elsewhere, but not now. Knowledge of the atmosphere and its behaviour had moved rapidly on, but so vast and complex was the subject, that without access to reams of daily observations, available only to centres like those in London, Paris and Berlin, even a competent judge like Maury could not hope to keep abreast of progress. 'Only in consequence of attentive and incessant study of atmospheric dynamics, as well as certain statistical facts . . . can adequate skills be attained in acting with these recently combined atoms of experimental knowledge.'

FitzRoy could rant against it all he wished, but he knew Maury's statement was dangerous. The name of Captain Matthew Maury still carried with it a commanding prestige among knowledgeable laymen, and for his opinion to be publicly featured in two widely read international scientific publications (for it appeared also in another French magazine, *Courrier des sciences*) could be extremely damaging, and FitzRoy knew it. 'In discouraging such forecasts as are now drawn in France and England, Captain Maury is unaware how completely he would destroy the scientific foundation of telegraphic *cautionary* notices.'

FitzRoy's reply to Maury's accusations was printed, in French, in both magazines. For good measure, he even wrote personally to Maury, but complaining did not help matters. Maury's reply when it came must only have incensed FitzRoy more, since he did not just repeat his ideas; he enlarged on them, in contradiction to everything FitzRoy had said.

While he was down, suffering under Le Verrier's snubs and Maury's blows, in came Marié-Davy with his own form of put-down. He wished to discontinue his correspondence with the British Meteorological Officer. Since language problems seemed to be the current fashion in excuses, he, too, resorted to blaming difficulty in translation – the lady who translated was accustomed to literary works and had problems with FitzRoy's meteorological language, he said, and he thought it better that the correspondence ceased, in case mistranslation led to misunderstanding. Although it was true that FitzRoy's densely scientific language and flowery metaphors did attract criticism even from native English speakers, the excuse still struck FitzRoy as being very thin. The most likely reason for Marié-Davy to cut contact with FitzRoy was on account of the excessively long hours he was being forced to work. He would have good reason to resent FitzRoy as being partly to blame.

Vaillant's vehement slandering of French forecasting was the root cause of Marié-Davy's troubles. Bolstered by the evidence he had acquired from FitzRoy himself, Vaillant suggested to Le Verrier that he ought to concentrate purely on notices of approaching storms, rather than attempting a blanket daily forecast. This was not an order, more of a warning shot, but Le Verrier was, like FitzRoy, suffering increasing isolation from his scientific peers on account of dabbling in the hocus-pocus of forecasting. And, like FitzRoy, too, his reaction was to strive harder to improve his accuracy. However, it was not Le Verrier himself who was prepared to undertake the striving – the onus for this he had delegated to the long-suffering Marié-Davy.

With Marié-Davy's support now gone, all hope of scientific endorsement for FitzRoy had effectively disappeared. In a world rapidly turning against him, he was beginning to rely more and more on his assistant, Babington. After the departure of the faithful Pattrickson in the wake of yet another Civil Service reshuffling exercise, Babington was now officially FitzRoy's deputy, central to the running of the department. Gradually, FitzRoy had educated Babington in the minutiae of compiling forecasts, and by 1864 Babington regularly constructed the everyday forecasts, the only person in the department, apart from FitzRoy himself, who could do so. Babington was also the one person in the world who not only understood FitzRoy's theories, but wholeheartedly supported them. When all others ridiculed his ideas, or refused to listen, Babington was there to approve, enthuse and assist. When there was frustration, he could share it. And it was only to Babington that FitzRoy could be open about his forecasting failures.

However, there was a problem in this. With Babington now producing his own independent forecasts from exactly the same figures, FitzRoy had an extra yardstick to judge his own performance against. If they both came up with the same interpretation, it was a satisfying justification of 'the method' – if it be right, then all to the good, and if it be wrong, then it was not their fault, but that the elements conspired to mislead them. When they differed, if FitzRoy was right and Babington wrong, then it was simply a case of the master's experience telling over the pupil's mistake, and FitzRoy could feel superior and act sympathetically. But when the time came that it was Babington who was right and FitzRoy who was wrong, then FitzRoy would not easily accept himself as second best – and this was already beginning to happen.

Forecasting the End

'This is the third time I have cautioned . . . and been wrong . . .,' noted FitzRoy to Babington one day in February 1864. 'I shall think less of uncertain (at present uncertain because not understood) lunar effects,' he declared.

A desperate man, recognising his own failures in forecasting, FitzRoy was prepared to abandon his very own pet theory, the theory of the effect of lunar and solar gravitational pull on the atmosphere, his lunisolar theory. The effect, he declared, appeared to counteract atmospheric conditions rather than enhance them. Babington, though, continued to calculate the lunisolar effect into his forecasts and was proved to be right when he predicted a storm approaching a few days later. Ignoring Babington's deductions, FitzRoy failed to caution and a ship was wrecked off the north-east coast. Sailors died in that storm and FitzRoy felt personally responsible for the loss of life. He abandoned restraint and immediately swung back to the other extreme. At the first signs of trouble, he was indiscriminately scattering warnings, while Babington's more specific forecasts were proved right, again, on the day.

'I am exceedingly vexed at the incautious and to a great extent determinedly perverse way in which I filled up the forecast table on Saturday,' FitzRoy chastised himself. 'We shall hear more from today's and tomorrow's Shipping Gazettes.'

Private admissions of inadequacy to Babington were one thing – it was a release to be able to discuss the reasons and correct his shortcomings. However, public admission of error was something entirely different. The pressure to be accurate was growing, and the more it grew, the more acutely he felt his failures. And all the while, Thomas Farrer was watching and waiting, looking for any excuse to knock FitzRoy from prominence and put him back where he had begun, behind a pile of ledgers collecting statistics. Milner Gibson's weather-comparison programme was his weapon, the test he anticipated FitzRoy would fail. Funding was another sword drawn against FitzRoy, constantly digging into his flank whenever he stepped out of line, which was frequently, since he had always found it impossible to operate up to his own standard of perfection within the confines of an inadequate budget.

As the weeks passed, with the department operating beyond its means and on borrowed time, the working atmosphere in the Office must have degenerated from purposeful to stressful, under a chief stretched to breaking point by the pressures placed upon him. As Darwin had found him in 1834, in 'a morbid depression of spirits' when he felt at the end of his tether, so must he have appeared to Babington. This time, thirty years older, health already suffering, and without a tight shipboard community to cocoon him, FitzRoy was facing the crux of a crisis with his inner reserves far shorter than they had been on board the *Beagle*. What would be the effect on his sanity if the Wrecks Department findings came down against him, and funding for his forecasts was removed?

The long-awaited report from the Board of Trade appeared in April 1864. Farrer, who would have been aware of the gist of it for weeks, must have found it difficult to contain himself until it was published. When at long last the printed copies arrived, Farrer sneered that maybe Scarborough was not a suitable place to use for reading the weather, since it seemed to disagree so remarkably from that which was expected of it. Leafing through the report, FitzRoy would have seen what he meant. Scarborough, where for some valid reason he had set storm cautions on several occasions, most of which had been wrong, was the one location that the Board of Trade – probably in the person of Farrer himself – had chosen to highlight as an example in its report. Not surprisingly, the report's verdict was that the level of inaccuracy was unacceptably high.

Following such a damning report, what were the chances of funding continuing? A few days later, the Meteorological Office vote came up for discussion before Parliament. Mr Augustus Smith, MP for Truro, was immediately on his feet. The expense of the Meteorological Department had risen dramatically since its founding in 1854, he decried. And why? Anything more absurd than the system by which FitzRoy and his staff professed to tell the weather was hardly possible to conceive, he said. Fortune telling, he went on, was punishable by law, but here was a government department foretelling what the weather would be two days hence. He said he had tested these forecasts by some results and found them about once right, twice wrong. He could not see why such meteorological observations as were needed should not be taken at Greenwich, where there was already a very economically run observatory. The French, he added, ran an institution in Paris of a much higher scientific order.

To be fair, Smith did not restrict his acidic attacks on expenditure to the Meteorological Department. Many others were on the receiving end of an equally vicious spume of contorted facts and logic. However, he did have a particular grievance where FitzRoy was concerned, since, as lessee of the Scilly Isles, his revenues had been curtailed as a result of some very pertinent gale warnings, which had much reduced the numbers of distressed vessels frequenting his harbours.

FitzRoy was not slow to retaliate and issued a detailed defence, which appeared next day in *The Times*. The establishment in Greenwich, he wrote, was not equipped to handle the nautical observations that the Board of Trade department had been set up to deal with; the increased expense since 1854 was mainly attributable to telegraphic communications, with forecasts themselves adding little to the costs; and the French institution he so praised received most of its information from the Meteorological Office in London. As for Smith's assertion about the incorrectness of prophecies, he referred him to the Meteorological Department's own report, recently issued.

Indeed, FitzRoy's own report of the Meteorological Department for 1864 had been published in April, just four days after the Board of Trade report on weather forecast accuracy. It was not so much an account of the work of the department and the progress made (because, apart from warnings and forecasts, little else was in progress these days) as FitzRoy's own justification of himself. Alluding to the Wrecks Department exercise, he denounced it by understatement. 'A separate official analysis and digest have been elaborately executed at the Board of Trade independent of this office. But I cannot quite admit the completeness of this exercise.' In retaliation, he put forward his own figures for the accuracy of his warnings, and for good measure quoted a letter published in the *Lynn Advertiser*. The writer, the Harbour Master at King's Lynn in Norfolk, had made his own analysis of all warnings in the area for the year 1863 and, in direct contrast to Mr Augustus Smith, had found them more or less accurate. And, to add weight to his testimony, he added:

> too much praise cannot be given to the author of the great boon bestowed upon the seamen and fishermen of the United Kingdom, and to all nations using our coast and harbours, for, glancing over the wrecks and casualties, numbering nearly 3000 for the year, they must strike with sorrow everyone who thinks of the happy homes made desolate. And there is no question but greater numbers would have been added, but for the timely warning given by Admiral FitzRoy.

The seafaring public, the very people he wanted to assist, were on his side, but they counted for little when men of stature were stacking up against him. However, one source did declare in his favour, and surprisingly, it was the editor of *The Times*. True, the editorial piece on 18 June 1864 started somewhat unpromisingly with a quote from Arago, 'Whatever may be the progress of the sciences never will observers who are trustworthy and careful of their reputation venture to foretell the state of the weather,' and went on to reflect that 'Admiral FitzRoy still has to convince the public, and at this task he labours yearly with most praiseworthy assiduity'. However, this *Times* leader-writer was prepared to believe the evidence of FitzRoy's own claims of accuracy in preference to the Wrecks Department's figures, offering the opinion that 'there can be no doubt that when Admiral FitzRoy telegraphs something or other is pretty sure to happen'. It is, though, the statement that concludes the piece that must have set FitzRoy's heart leaping.

> The advantage of such a science [i.e. weather forecasting], if it could be founded, would be so incomparably greater than any pecuniary loss that we are likely to incur in the attempt, that we heartily wish him success, and are disposed to make every allowance for the special causes of occasional failure which have sometimes brought discredit on his prophecies.

If only the government would take such a view – and, amazingly, it did. The money was forthcoming, storm

warnings and forecasts continued. But the fight, the suspense
and the constant struggle to justify himself had exhausted
FitzRoy. To cap it all, his battle for fame as a meteorological
theorist was lost. The only foreigner to stand by him was
Mouchez of the Ministry of Marine in Paris. But Mouchez
was, like himself, sidelined by scientists, Marié-Davy and Le
Verrier, for example, regarding him as totally removed from
the circle of the elite. Mouchez tried hard to obtain for FitzRoy
the honour he felt he deserved in recognition of his
contributions to French meteorology, but the best he could
manage was the presentation of a travelling clock from his
own ministry. No doubt FitzRoy accepted the gift with
gratitude and grace, but the sincerely intended accolade
looked more like an affront when compared with the Legion of
Honour awarded to Maury by the French government a couple
of years earlier. FitzRoy must have felt like throwing
Mouchez's clock back in his face.

If FitzRoy chose now to retreat from the fray and abandon
himself to his fate; if he decided, for the sake of an easy life, to
retire into his departmental work, collect statistics, issue
warnings and take home his salary to his wife, then who could
blame him? He was, after all, a man in his sixtieth year, worn
out by failure, frustration and stress. A naval officer, however,
was trained not to surrender while there was a battle still to be
fought. A reprieve for the storm warnings might have been
granted, but their future was not secure. The seafarers and
fishermen whose lives he set out to save still needed him. But
the opposition, whose ranks were led by Francis Galton, James
Glaisher and Thomas Farrer, was building in strength. All it
needed was for him to put just one foot wrong, and they would
overrun his department, reverting it to a statistical sweatshop
geared simply to profit their own ends. If he allowed this to
happen, then he would be letting down everyone who
depended on him – his mariners, his staff (especially
Babington), his family and himself.

So he threw all his efforts into improving and perfecting his
methods of forecasting. He drew up a list of pointers to be
followed, the indications of approaching weather, how to
detect an impending storm. He worked on his graphs of
temperature and pressure, swapping and changing the format
until he found the best way to give the quickest warning of
large-scale changes. He discussed at weary length with

Babington the optimum timing for issuing storm cautions, castigating himself on many an occasion for not giving warning soon enough and thereby endangering lives. Babington tended to opt for later, better-targeted warnings, seeing unnecessary cautioning as leading to the inaccuracy figures they were fighting to avoid. It was a question of 'crying wolf' – the danger in issuing too many warnings that were too far in advance, or which came to nothing, would be that eventually all warnings would be ignored. Better to miss some gales than to warn where there was nothing to be wary of. Signals, he advised, should be reserved for great gales.

Babington might have been able to limit FitzRoy's extravagance in the issuing of warnings, but he could do nothing to curb his spending. Lack of regard for budgets had been a feature of FitzRoy's life. He had never learned to trim his outlay to match his means, a trait no doubt attributable to his privileged upbringing, where money had been no object. By the end of 1864, the money was running out, again. The decree to reduce spending came this time directly from the Treasury. The Board of Trade had long ago lost any confidence in FitzRoy's abilities to make stringent cuts, so it swung the axe for itself. For the chop were eight of FitzRoy's most valuable telegraph outposts in Devon, Pembroke, Scotland and Ireland. Messages to and from these stations ceased with immediate effect.

For FitzRoy, a vital part of the anatomy of his network had been amputated. Without the lifeblood of information flowing from the extremities of his territory to the hub of his London office, his ability to diagnose the symptoms of the weather and pump back warnings was severely restricted. It meant his desperate bid to improve accuracy could now only ever be a dream. With curtailed observations, he would need to run faster just to maintain the quality of service he was already providing. Just as an injured body decides to withhold support from its extremities in order to preserve life in its central core, so FitzRoy began the process of withdrawing from all his activities apart from the key function of providing storm warnings. By February, the final line had been drawn under the simultaneous-observations exercise, when Babington brought to him his concluding report. Even the lunisolar observations, which were still being assiduously collected and collated by Babington, fell into neglect.

The effect on FitzRoy's health was devastating. Two years earlier, he had already been complaining to the Board of Trade of the way stress was weakening his health. A year later, his hearing was deteriorating to the extent that he had admitted to Babington that he had not been able to hear the wind, 'being hard of hearing at any time'. By the beginning of 1865, not only were his ears failing, but his mind was in decline, too. Babington would later comment that 'his naturally clear and vigorous mind – clouded and weakened by the continuous strain upon it and the anxiety he had undergone – was incapable of sustained exertion.'

By covering up his mistakes and keeping a controlling hand on the work of the Office, Babington managed to conceal his chief's mental state from their superiors at the Board of Trade. However, come March, FitzRoy could hide his ever more frequent absences no more, and was admitting to the Board, and consequently to himself, that his health was failing. Finally, FitzRoy decided that perhaps it was time to heed the advice of his doctor, to detach himself from his worries and to stay at home and rest. Once the decision was taken, he let his mind relax and settled into tranquillity.

However, this peace of mind did not last long – it crumbled before the first day was out with the arrival of a letter. Signed Revd F.W. Tremlett, the note informed the admiral that his friend Captain Matthew Maury was temporarily lodging with him in London, and invited Admiral and Mrs FitzRoy to come and stay from Saturday, 29 April, for two nights, as the captain wished to see the admiral before departing for the West Indies. Maria, his wife, found him pacing the room in an agony of indecision. To go might heal the rift between himself and Maury – after all, why else would Maury be so desirous of seeing him, and for so long a visit? The thought of easing just one of his torments was tempting. On the other hand, the effort of stretching his tired mind in debate with Maury was daunting. The vow he had made to rest and relax was only hours old, and should not be broken. He decided to refuse the invitation, and had Maria send Tremlett a note to that effect. But come Saturday morning, he was regretting the refusal. Declining Maria's suggestion of accompanying her on a drive, he waited until she had disappeared in the carriage, rushed to the station, caught a train into London, and thence a cab out to Belsize Park, where he alighted at the home of Revd F.W. Tremlett.

Maury and FitzRoy had not met for some time – quite possibly their paths had not crossed at all during the years since 1862 when Maury had returned to England. They each must have regarded the appearance of the other with a degree of shock. FitzRoy's illness and overwork had aged him in the past two years, and the once robust Maury was bowed and greyed, also through ill health. Maury had based himself and his young son Matthew in the small Cheshire town of Bowdon. Handy for Manchester, where he could acquire the knowledge and materials of explosives, it was also close to the school at Rose Hill founded by the former tutor of his son John, and where he had enrolled Matthew. Unfortunately, his notion of nurturing his son in the peaceful atmosphere of the English countryside, well away from war, had gone horribly wrong when he had been taken ill. Matthew junior had been forced to transform himself from schoolboy to nurse, a role that could not have sat too easily on the shoulders of a 16-year-old, guiding his father through the trials of nineteenth-century surgery and recuperation. But now Maury, recovered, had been ordered by the Confederates to leave the peace of England to return to war-ravaged America. He and his son had alighted for a few days with the Revd Tremlett before setting sail from London for the West Indies, from where they hoped, somehow, to secure a passage home to Virginia. In advance, with his boxes and trunks, he had dispatched the casks of explosives and other materials he had been sent to obtain, while packed into his brain was the armoury of design improvements guaranteed to transform his torpedo into an effective weapon. He could be returning as the saviour of the Confederacy – if it was not already too late.

No doubt the greeting exchanged by the two old friends was cordial. Maury would not have expressed a desire to meet and socialise with FitzRoy, especially in the company of his wife, had he not wished to mend matters and take his farewell in friendship. FitzRoy, in his delicate state of health, could surely not have intended to force a showdown. Despite an aristocratic self-assurance often interpreted as arrogance, outright aggression had never been his nature. He was a humanist at heart, and his sympathies must have been with Maury, heading off into an uncertain future, not knowing the straits in which he would find his family – if indeed he would ever find them at all. Intentions, then, on both sides must have been peaceful and reconciliatory.

But peaceful and reconciliatory the reality certainly was not. Words were exchanged or actions taken that Saturday afternoon that were so explosive they propelled FitzRoy from Revd Tremlett's house and all the way back to his home in Norwood. At 8 p.m. he staggered into the house agitated and fatigued in the extreme. Whatever the demon that had been loosed, it would not let him rest. It worried at him and plagued him so that, late that evening, he announced to his wife he would go back next morning for the return round with Maury. Maria was astounded. Another confrontation must surely tip his teetering balance of mind into insanity. Talk about it in the morning, she gently advised. It was a conversation they would never have.

That Sunday morning, after a quiet night, FitzRoy rose early, kissed his young daughter, Laura, still asleep in her bed, went to his dressing room, picked up his razor – and cut his own throat.

The Times, Tuesday May 2nd, 1865:
Yesterday morning a painful feeling of regret agitated the whole of the officials of the Board of Trade, on assembling at Whitehall, when the melancholy news of the suicide of Vice-Admiral FitzRoy, the chief of the meteorological division of the Government department, became known. He cut his throat at his residence, Lyndhurst House, Norwood, on Sunday morning.

Why did he do it? There is no means of knowing exactly what went through his mind in the early hours of the last day of April 1865. There are, though, various theories. One is the deep theological divide between himself and Darwin over the theory of evolution. Then there was his desperate bid to eclipse Darwin's fame with his own. Not a gambling man, he did wager all on his lunisolar theory, and lost. Another version holds that he cut his throat because he had become a public laughing stock over his forecasting failures. The evidence of his reaction proves that each of these deeply troubled him. Cumulatively, the effect on an unbalanced psyche was ruining his physical and mental health. Whatever the individual theories, though, the deed itself was triggered by the meeting with Maury. What occurred in the Revd Tremlett's drawing room, according to FitzRoy's wife, was 'a circumstance which had impressed him awfully', though what it was that had awed him we do not know. Perhaps Maury did not know either. It is difficult to imagine that Maury, who had always supported and praised FitzRoy in the past, should suddenly seek to find fault. FitzRoy was upset by his comments about the inadvisability of forecasting in Britain's island situation, but perhaps, had he not spun off into a purple passion at the first sniff of criticism, he might have taken time to consider Maury's statement and have found good common sense in his reasoning.

In America, both Maury and Henry had operated on the east coast, backed by a barrage of regular, reliable observations transmitted from the inner continent from whence the prevailing weather came. They had both used these observations to predict the coming east-coast weather, with some success. FitzRoy was attempting to achieve the same rate of success in Britain without any knowledge of the approaching weather. Imagine a weather chart with no line of depressions marching in from the Atlantic, no isobars, no rainfall clusters, no wind arrows, no cloud pattern, just a blank expanse over all points west of the coast of Ireland. FitzRoy's charts were just like that – he had to guess what filled the blank, constructing a picture from the little he knew. He maintained that, with careful calculation, the barometer, the thermometer, the hydrometer and the anemometer told the experienced observer, the state not of the current weather, but of the weather to come. There is some justification in this. Just as a skilled archaeological specialist can reconstruct a complete Roman amphora from the evidence of a sherd of pot, so the formation of an entire cyclonic depression might be gleaned from a glimpse of its rim. But the archaeologist needs a database of all known examples of whole amphora in order to reconstruct accurately. The equivalent database in terms of climatic phenomena still cannot be accurately modelled despite today's gigantic number-crunching computers, and certainly did not exist in FitzRoy's day. Galton was right – only with extensive research could more knowledge be obtained. Perhaps Maury tried to point this out to him, in the hope of saving his sanity. Perhaps he told him that, with the limited tools he had, he could not hope to improve his accuracy beyond that which he was already achieving – and that perhaps, maybe, daily forecasts, as opposed to occasional storm warnings, would be better left unpublished.

It would not have been what FitzRoy wanted to hear. If his forecasts never could be made accurate, then what was the point of continuing? And if he stopped forecasting, if he abandoned his last defence, the way was open for the invaders – Galton, Glaisher and Farrer, plus the collective ranks of the British Association, the Meteorological Society and the Royal Society – to march in and erect their sterile statistical service on the discredited remains of his own. By cutting his throat, he made sure he would never live to see that day come.

On Tuesday, 2 May, the day the report of FitzRoy's suicide appeared in *The Times*, Maury set sail for the West Indies, taking with him the only eyewitness evidence of exactly what had taken place that fateful afternoon in Belsize Park. If he did know the direct cause of FitzRoy's fatal unease, then he was not saying. Babington, as FitzRoy's deputy, assumed responsibility for the day-to-day functioning of the department he was already running in all but name. Daily weather reports, plus two-day forecasts and cautionary storm warnings, were issued – and with much the same measure of accuracy – as if nothing had happened.

But something was happening. Thomas Farrer, for one, was immediately active. With FitzRoy out of the way, the opportunity was suddenly there to get rid of the pseudo-scientific claptrap permeating the Meteorological Office and dismantle the whole paraphernalia of weather forecasting, which brought derision down onto the good name of the Board of Trade. Of course it was not in his power to do this himself – the axe had to be wielded by his political master, the President of the Board of Trade, Milner Gibson. And Milner Gibson needed persuading, for, whatever Farrer's thoughts on the matter, the fact remained that storm warnings were considered valuable by many voters, and even the weather forecasts themselves were popular with the public. So the Board of Trade did what it always did when faced with a decision on a scientific topic – it referred the whole matter to the Royal Society, and awaited its report.

Mr Augustus Smith, MP, was not prepared to wait. Like Farrer, he scented the chance to eliminate storm warnings, the charitable reason being their cost to the Exchequer, which, as self-appointed spending watchdog, he had always been at pains to point out. However, there was also the advantage to his personal purse to think of – with storm warnings gone, the increase in the numbers of storm-battered ships staggering into Scilly Isles ports would restore some of his lost revenue. The flowers had hardly faded from FitzRoy's funeral cortege before Smith was up on his feet in the Commons. *The Times* reported:

> He wished to know what was the intention of the Government as to filling up the vacant post in the Meteorological Department, as some returns laid before the House last year showed the weather prophecies were twice wrong for once right. On last Tuesday week a tremendous gale blew over the northern part of the kingdom, yet the Board of Trade promised for that day weather fresh and moderate and prophesied a direction of wind different from that which actually prevailed. Almost all great storms came on before the signal was shown, and on Tuesday week last the cautionary signals were not exhibited before the gale had begun.

Milner Gibson replied with a spirited defence of FitzRoy, a defence that would have gladdened the admiral, if only he had not had to die before it could be spoken. 'In the opinion of those best qualified to judge, the deceased Admiral did great service in perfecting the science of meteorology.' He also quashed one argument of the FitzRoy detractors with the assurance that his departure into the practical application of meteorological science, including weather prediction, 'had been done by the late Admiral FitzRoy, very much with the concurrence of that House and of the Royal Society'. However, he did add that

> Since that Admiral's lamented death, he had felt right, seeing the original duties of the department had been very much abandoned, again to seek the advice of the Royal Society and to ask them whether the science of meteorology was in so perfect a condition as to justify the Government in proposing votes of money for continuing the weather forecasts.

The Royal Society sprang into action by appointing a committee. Its members were Staff Commander Evans of the Hydrographer's Office (nominated by the Admiralty), Thomas Farrer himself, and, as chairman, Francis Galton. Just as FitzRoy might have feared, the men who had pressurised him most by their concerted opposition to him were the very people appointed to sit in judgement over him. The doors of FitzRoy's office were opened to them and they rushed in with a relish, poking, prying and delving into his books and his records, questioning his staff and calling in evidence from external experts. They had a mission to fulfil, which they attacked with the full vigour of their prejudices. Impartiality would feature thinly in their deliberations.

Galton's own hostile attitude was well known. He went into the investigation with the belief that forecasting the weather was an impossible act undertaken only by fools and came out the other end with much the same opinion: 'We can find no competent meteorologist who believes the science to be at present in such a state as to enable day by day weather to be forecast for the next 48 hours.'

In his *Meteorographica* he had rebuked FitzRoy for his backlog of unpublished data and there it was, 550,000 sets of observations, most of which lay just as they had been received, untabulated and uncollated. It irritated Galton to think of them just sitting there gathering dust, while he and other scientific researchers had been forced to go out and seek their own data, and he concluded that the problem was entirely of FitzRoy's making. Next, he went to unearth FitzRoy's collation books, those books of tabulated figures whose function and format the Admiral had agonised over, ten years before. These Galton disliked immediately – too much raw data had disappeared in the abstractions. And FitzRoy's popular and highly successful wind-star charts he rejected in disgust. He had never liked them, because they submerged hard data into a purely graphical form; he ignored the fact that navigators loved them for that very reason.

Finally, the entire weather-forecasting operation was exposed to the full glare of the committee's scrutiny. Babington was subjected to intense questioning on the method FitzRoy had laid down for forecasting – if indeed it did exist at all. Of course there was a method, retorted Babington, otherwise how else would the two of them have been able to arrive at the same forecast given the same variables? And he tried his best to explain it. But, naturally, nothing existed in the form that a set of scientists and statisticians would expect to see. There were no lists of given maxims, no formulae, no calculated probabilities. It was a constantly evolving process that shifted and refined itself based on experience, tempered by precedents proved and disproved.

There remained the problem of physically where, how and by whom the meteorological service should be carried out. It could stay where it was, under the Board of Trade, whose official powers could be brought to bear to ensure the cooperation of the marine and naval fleets, or it could be removed from government altogether, to a scientific body better qualified to interpret and present the results. Greenwich and Kew observatories, and even the Meteorological Society, with James Glaisher to oversee proceedings, were considered.

In April 1866, nearly a year after FitzRoy's death, the decisions had been made and the report was ready to be presented to Parliament. As might be expected, it was utterly damning of FitzRoy and his department, with Francis Galton's hand evident in the demolition, right from page one. Its major recommendation was that there should still be a meteorological service, since its existence was provided for in the terms set down after the Brussels Conference, and that the main focus of the service should return to those terms – that meteorological observations should be collected at sea, collated in a central office and the results made available to all who might benefit, including scientists. The problem of physical organisation was solved by the suggestion of establishing a split service – a government office to oversee the distribution of instruments and the collection of observations, and a scientific body to tabulate and digest the figures. To avoid a repeat of the FitzRoy scenario, where one man had the power to direct the operation along his own idiosyncratic lines, the two offices should be overseen by a committee under the auspices of the Royal Society.

The report also discussed forecasts and the current method of predictive storm warnings. Since the erratically inaccurate nature of daily weather forecasts threatened to prejudice public opinion against the entire science of meteorology, the recommendation was that they should be abandoned. Storm warnings, though, were not such a clear-cut case. The naval man, Captain Evans, could be expected to understand the benefit of them; Farrer, who until recently would have given anything to see the back of them, appeared to have changed his mind – perhaps public campaigning and political pressure were persuading him to find in favour of them; Galton, surprisingly, voted in support, too. No doubt as a traveller himself, and one who had entrusted his fate to the vagaries of the ocean on long sea voyages, he was aware of the need for them. Trawling with Babington through FitzRoy's methods, he had understood the admiral's rationale and realised that a system for predicting extremes of weather could, to a certain extent, be codified, and therefore could come legitimately under the province of a scientific body. Storm warnings, therefore, could stay.

Babington was outraged by the findings of the committee. If the office were to be divided, then he could foresee all the difficult, thankless tasks being left at the Board of Trade, while the progressive work, the published results and all the kudos would emanate from Kew, or whichever scientific body were chosen. As far as he was concerned, the existing staff were perfectly capable of doing all the work, whether it be for the Board of Trade or a Kew committee, and he thought their zeal and enthusiasm should be rewarded by something better than relegation to the dogsbodies of the operation. He withdrew his application for appointment as FitzRoy's successor, and threatened to resign.

Farrer used the threat to have daily weather forecasts discontinued immediately, without waiting for ratification of the recommendations of the report from Parliament. He then sounded out Babington's assistant, Simmonds, on taking over the storm warnings. Simmonds refused, saying that, although he knew how to predict the storms, there was no way he was going to take responsibility for them – and who could blame him at such a sensitive time? Farrer found himself in a quandary, for, without Babington, he could not continue to issue the storm warnings he was now convinced he needed to keep. He even prepared, ready for issue, a circular on the Interruption of Storm Warnings. Fortunately for him, he was saved by Babington, who appears at this point to have postponed his resignation. Fiercely loyal to his deceased master, Babington no doubt realised that, without him to continue them, FitzRoy's system of storm warnings would die with him. It could be only a temporary reprieve for Farrer, though, since he still needed the government to take a decision on the future of the department and, more vitally, vote the necessary funding for the warnings. He decided to enlist Milner Gibson's assistance in pushing for action. The outcome was not what he would have been hoping for, and once more it was the man at the helm of the Imperial Observatory in Paris, Le Verrier, who was to blame.

Le Verrier himself was not without his problems, since in France, too, predictive storm warnings were being swallowed into storms of their own. Like FitzRoy, Marié-Davy had buckled under the pressure of work. A complaint to Le Verrier that his health was suffering brought not a reduction of his duties but dismissal from his post. However, Le Verrier soon discovered the service could not function properly without him, and asked him to return. Not surprisingly, Marié-Davy refused to come back to the status quo and demanded concessions. Le Verrier responded not by acquiescing but by terminating the storm-warning service in October 1865, six months after FitzRoy had died, and while the future of the British operation was being investigated.

With his warnings, forecasts went too, disappearing from the *Bulletin international* and other published outlets. The forecasts themselves were not much missed. In trying not to get them wrong, the French meteorological service had erred on the side of vagueness. Forecasting general weather rather than exceptional weather was proving difficult elsewhere, too. Even the celebrated Dové in Berlin had ceased to provide a daily weather forecast. However, Le Verrier was not happy to abandon storm warnings completely, in case the Ministry of Marine decided to turn to the British again, so he installed an immediate replacement. Every day he sent a report of current weather to all ports, from which they could draw their own conclusions. It was the system he had once tried to force on FitzRoy, a system he himself derided as having no element of science attached to it whatsoever, but he preferred second best to being held to ransom by a man he disliked.

Anxiously, though, he looked across the Channel to developments there. FitzRoy's superior system was still being operated with equal success by Babington, making the French system appear decidedly mediocre in comparison. But there were strong rumours that the Royal Society's Committee of Enquiry would act to quash storm warnings altogether. He must have been praying for a means to urge just such an outcome when the ideal opportunity landed on his desk. Sabine had written, wanting to know the method he used to forecast storms in France. He composed a detailed report describing his current system of transmitting only facts of actual weather and emphasising that he no longer made any attempt to predict. Unfortunately for him, he was too slow. His report arrived back in Britain after the committee had reported in favour of FitzRoy's system of storm warnings. However, he was not too late to influence the thinking of the man who now held sway over the fate of British storm warnings – Milner Gibson, President of the Board of Trade.

Farrer received Le Verrier's report, and cheered at its contents. Here was a man operating a perfectly good storm-warning system that did not require any science for its management. An arrangement like this would solve Farrer's problems at a stroke. He would not be reliant on Babington – anyone could be brought in to operate it – and he would still have a storm-warning system. He sent Le Verrier's report to the President, emphasising that Le Verrier

> abandons the notion of giving any general mention of weather for two or three days beforehand – as Admiral FitzRoy supposed *he* could do; and that as regards storms dangerous to Sailors he does not profess to give absolute direction, or to announce bad weather more than 24 or sometimes 12 hours beforehand.

He had not, though, calculated on Milner Gibson's reaction.

> Very interesting – but after all, there is not much practical utility in foretelling storms if the warning precedes the storm by so short a time as twelve hours – the people of the locality with the use of a barometer – can generally form a pretty good idea of the weather a few hours before a gale begins.

With those words, Milner Gibson dismissed Le Verrier's system. And, what is more, he recoiled from accepting the committee's recommendation to maintain warnings as they stood. Storm 'prophecies' smacked of quackery and he and the board would not be seen to be funding anything remotely 'scientific' without further backing – from where else but the Royal Society. Farrer was trapped by the very weapon he had himself used against FitzRoy. And, like FitzRoy in the same situation, he appealed to Colonel Sabine for his support – a scientific testimony for the continuing of storm warnings.

He was wasting his time. The reply from the Royal Society, in October, gave backing to the bulk of the Committee of Enquiry's recommendations – the split of duties between the government (for the collection of data) and the Royal Society (for the 'digestion and tabulation of results'), government funding of the Royal Society for the purpose, with the clearing of the backlog of accumulated observations as a priority. However, it did not endorse the continuation of storm warnings.

> At present, these warnings are founded on rules mainly empirical. In a few years they may probably be much improved by deductions from the observations in land meteorology, which will by that time have been collected and studied. The empirical character may thus be expected to give way to one more strictly scientific, in which case storm warnings might fitly be undertaken by a strictly scientific body.

Farrer was shocked. He knew that, without the Royal Society's backing, there was little chance of obtaining funds to continue the storm warnings. He reminded Sabine that, without storm warnings, meteorology would instantly lose its popular appeal and the parliamentary vote they were looking for might well not be forthcoming from a House that saw no instant return for its money. 'Am I not right in supposing that to omit storm warnings will strike £3000 off our estimate?'

Inevitably, with the Royal Society against them, and without a fighter like FitzRoy to take on the personal responsibility for them, warnings would have to cease. On 29 November 1866 the Board of Trade issued a circular announcing that cautionary storm warnings would be discontinued as from 7 December. On that day Babington resigned, and storm warnings looked set to join weather forecasts as a piece of history.

T W E N T Y

Rebirth

On the wintry morning of 1 December 1866, the Revd Redford, Vicar of St Pauls in the Cumbrian port of Silloth, first received the news of the suspension of gale warnings. Perhaps it was rushed to him by the harbour master of that Cumbrian port, horrified at the news, for Silloth, positioned on the extreme north-west corner of England's coastline, bore the brunt of the worst of the westerlies rushing over the Irish Sea. The reverend was himself a Fellow of the Royal Society, and, aghast at what had been perpetrated by the President and Council in the Society's name, he took out his pen and wrote immediately.

> I trust the suspension will not be of long continuance, for the '*warnings*' are *invaluable* on this coast, and I think that if the President and Council of the Royal Society could have witnessed the growing attention paid to the Signals by the sailors and were aware of their general accuracy they would not have recommended even a temporary suspension.

The Revd Redford's protest was the first of an avalanche, arriving at the door of the Board of Trade by nearly every post. Dundee Harbour Trustees were also quick off the mark. They recognised that the system might well be improved by a better understanding of science, 'but surely this furnishes no valid reason for suspending the present signals?' They urged the Board of Trade to consider 'the extreme disappointment and danger which will arise from the discontinuance of storm signals, more especially at this season of the year, when sudden and violent changes of atmosphere and consequent storms may be anticipated'.

Indeed, the Board of Trade showed little regard for mariners, choosing to suspend the gale warnings just before the onset of the winter gales and, for some, this was just too much. At the end of February came a petition signed by all the shipmasters, first and second mates and chief engineers of the steamers and sailing ships in the Port of Leith. They had, they said, 'set a high value on the system of cautionary warnings adopted by the late Admiral FitzRoy' and they considered suspension 'a step disadvantageous to the mercantile marine, especially when viewed in connection with the fearful and lamentably destructive gales which visited the country in the January and February of this year'.

In Dundee again, the Local Marine Board was not so much baffled by the timing, as by the reasons. 'The Local Board would not be understood as expressing any opinion on the scientific questions involved in the law of storms, or as to the rules on which storm warnings were founded . . .' but it did know 'that in the majority of cases the warnings have been useful, and therefore, although not in every case verified by results, the continuance of them would be a benefit. The Local Board view the subject as one of deep importance to the mercantile marine interest.'

It was not just the shipowners and sailors who put pen to paper. Maritime insurance underwriters were equally concerned. Liverpool Underwriters were 'much in favour of the cautionary signals', while those in Glasgow and Greenock protested that, 'however imperfect the "Storm Signals" were . . . [they] considered them well worthy of all the money expended in carrying them out, as the caution thereby produced was calculated to save many lives and valuable property'. And the Manchester Chamber of Commerce revealed a quite unsuspected beneficiary of the warnings: 'in one of our largest collieries in this district they have been of great use as indications of atmospheric changes. For such purposes as well as nautical, it is felt very desirable that they should be resumed.'

Yet perhaps the most influential and authoritative support of storm warnings came from someone who was no layman, no humble sailor, no petty official, but a highly respected meteorologist in his own right. He was C. Piazzi Smyth, Astronomer Royal for Scotland. Piazzi Smyth alone among scientists was ready to point out a truth that to him was graphically obvious.

Now, it is quite true, that if the results be examined in a hypercritical manner, and according to the usual ideas connected with the exact science of the highest order, they will by no means come up to the mark. But then, high science is one thing, and storm warnings so completely another that it is not fair to measure its use and right to existence by a test derived from anything else of so entirely a different nature . . .

Storm prediction happily does not depend upon meteorological science being improved up to be like astronomy . . . (a consummation not likely to occur within the limits of the duration of the coalfields of this country).

In the midst of this mountain of correspondence came one letter, in French, directed to Babington himself. M. Le Verrier was wanting to know exactly what was going on. If forecasts and storm warnings were to disappear, what was to happen to the exchange of telegraphed weather reports on which his own service relied? Le Verrier had cause to be worried, and the irony is that it was all of his own making.

Despite his report to Sabine, he himself was totally dissatisfied with simply transmitting weather reports. He needed to restore a more satisfactory service, but who could provide it? The feud between himself and Marié-Davy worsened, with accusations of bad management and incompetence flying in both directions. Marié-Davy, in despair, even sent an entreaty to the Emperor, attempting to blacken the name of Le Verrier, the Emperor's favourite, which did absolutely nothing to improve the position. Instead, Le Verrier became even more determined to prove that a proper storm-warning service could operate efficiently without Marié-Davy. In May 1866, just a month after he had sent his report that effectively ensured the demise of British forecasts and storm warnings, he began his own warnings again, along with a forecast of when bad weather would start, and how long it would last. With the buffer of their ex-chief gone, Marié-Davy's staff were driven to work long hours every day under the lash of Le Verrier, an impossible taskmaster. He knew his position was shaky, that his staff worked under pressure and duress. If the telegraphed reports from London were to cease, accuracy would suffer, and his new system would be heading for total collapse.

In London, meanwhile, no reinstatement of storm warnings looked likely to materialise. All the letters, memorials and petitions of protest had been passed on by Farrer from the Board of Trade to the Royal Society's Meteorological

Committee, where they had disappeared into a black hole of indifference. Public outrage finally came to a head in the office of Charles Henry Lennox, Duke of Richmond, the successor to Milner Gibson as President of the Board of Trade.

> A large deputation has waited on the Duke of Richmond to urge that some warning should be given of apprehended danger from storms, and I am asked whether it might be possible . . . to give effect to a desire which is strongly expressed by many competent and influential bodies . . . [and] whether it might not be possible for the Committee appointed by the Royal Society, upon such conditions and under such limitations as they may think fit . . . [to give] some warning of apprehended danger from storms.

The officials of the Board of Trade seemed to have conveniently overlooked the fact that, technically, it was they who had refused to endorse funding for the Royal Society's Meteorological Committee to continue storm warnings – based on the Royal Society's own recommendation, it is true. Funding and responsibility for warnings, forecasts and meteorology in general had passed back and forth in an intricate game of brinkmanship between government departments and scientific bodies ever since the Brussels Conference. Who should have the Meteorological Office, who should fund it and even who should appoint – and be – its chief had been under dispute at the start. An awkward alliance between the Admiralty and the Board of Trade eventually, in 1854, gave birth to the department, with FitzRoy's appointment being made by the Admiralty, although his duty manager was Beechey of the Marine Department of the Board of Trade, and his job description was supplied by the Royal Society.

The Board of Trade never did appear to be comfortable harbouring this quasi-scientific beast in its bosom, especially when its chief began to veer off at a tangent, dabbling in meteorological theory. Farrer and various Presidents itched to bring him back to being merely the collector and tabulator of observations – but they always felt that somehow his activities were approved and directed by the Royal Society. It was not until after FitzRoy's death that Farrer realised this was not the case. When Milner Gibson declared in Parliament that

FitzRoy's weather predictions were condoned by the Royal Society, Sabine was at pains to point out that 'it must not be forgotten that storm warnings did not originate in any recommendation of the Royal Society'. Farrer's response was that he wished that the society could have made this clearer earlier, so that 'we might have prevented and altered much in the Proceedings of the Meteorological Department, which with the doubtful utterance and semi-approval of the Royal Society, we laymen were unable to touch'.

The problem was that the Royal Society rarely did make its meaning very clear. Back in 1852, a report written by Sabine had led Maury to believe that the Admiralty had backed his conference proposal, when indeed it had not. In 1854 the society's instructions for the function of the department, again emanating from Sabine, were couched as recommendations, allowing FitzRoy to treat them exactly as that, and leaving him free to decide whether to follow them or not. He chose not. The same happened again when FitzRoy was fighting desperately for his predictive storm warnings. The Royal Society declared it could not be seen to be certifying his methods as scientific, but it did not actually state that he should not be permitted to undertake what he proposed. That decision was left hanging, nobody took it, so FitzRoy went ahead. The Board of Trade, interpreting the lack of express disapproval by the Royal Society as actual tacit approval, was happy, provided FitzRoy shouldered the responsibility himself.

After his death, and Babington's resignation, when the whole question of storm warnings surfaced yet again, the Royal Society refused to take on the task. This time there was no FitzRoy, and no one like him within the government capable or sufficiently motivated to undertake a storm-warning system. So warnings disappeared along with daily forecasts, and with them went the £3,000 in funding, just as Farrer had warned.

With the hue and cry of public protest and with the weight of government pressure bearing down on them, the members of the Meteorological Committee of the Royal Society had no option but to bow to popular opinion. Sabine, as the society's President, tried to hide under technicalities, in order to avoid blame. In a letter to Farrer in April 1867, he said he wished to make it clear that it was a Board of Trade decision to stop storm warnings, the Meteorological Committee having 'declined' to request it to do so. On the following day, he was complaining that the Treasury seemed to be under the impression that the Royal Society was an executive department of government, undertaking meteorological research. The Royal Society was not a government department, he was at pains to point out. All the research was undertaken under the auspices of a committee appointed by it. However, no matter how the Royal Society tried to wriggle out of responsibility, its Meteorological Committee had to be seen to be taking action towards the restoration of storm warnings.

The committee, with Francis Galton still prominent among its members, had met officially for the first time on 3 January 1867. Its first action had been to define the structure of the new Meteorological Office and appoint its staff. Captain Henry Toynbee of the Mercantile Marine became Superintendent of the Marine branch. His job mirrored FitzRoy's original 1855 brief to distribute instruments to ships and collect observations, which was precisely the post Babington did not want to fill. Secretary to the Committee and Director of the Kew Observatory was Mr Balfour Stewart. Kew was where the true business of meteorology was to be conducted, where instruments were calibrated, land observations collected and results collated and analysed. In overall charge of both branches, and FitzRoy's true successor, was Robert Henry Scott, the Irish meteorologist who had translated Henri Dové's *Law of Storms* and had thus provided FitzRoy with the fodder for his scientific ideas.

Scott, like FitzRoy before him, came to the Meteorological Department as the personal protégé of Colonel Sabine. The son of a prominent Dublin lawyer, Robert Scott had studied at Trinity College in Dublin, graduating at the age of 22 as Senior Moderator in Experimental Physics. From there he had gone to Germany for two years to further his scientific studies. Although these included meteorology under Dové, his speciality seems to have been mineralogy, for he was working as Lecturer in Mineralogy to the Royal Dublin Society when Sabine, a Dubliner himself and friend of the Scott family, approached him for the post. Unlike the furore that had accompanied FitzRoy's appointment, Scott appears to have been the sole contender – perhaps because FitzRoy's dismal end pointed to a jinx on the job, or more likely because Sabine's opinion, as President of the Royal Society and member of the Meteorological Committee, was virtually unassailable. Although Scott did have a scientific background, and some knowledge of meteorological theory, he was not renowned for the thrust of his ideas. He was, though, thorough and meticulous in his application to his work and, in contrast to FitzRoy, thrived on routine and order. No doubt Sabine hoped to succeed where he had failed with FitzRoy in selecting a biddable character for the job this time round.

Staff in place, Galton, Sabine and the other members of the committee pocketed the £10,570 voted to them by Parliament, and straight away set to work to implement big ideas. Almost immediately, they came across FitzRoy's foremost problem – the amount of the vote was not sufficient to pay for everything they wanted to do. The first casualty was a planned network of seven land observatories with state-of-the-art self-recording instruments, which had to be slow-pedalled after the construction of only three machines. They launched with great gusto into the trial of the first telegraphed transatlantic reports from Heart's Content, Newfoundland, only to find the expense of this meant they were forced to abandon plans to reinstate those remote telegraphic outposts in Wales, Scotland and Ireland that had been arbitrarily cut as a result of FitzRoy's overspending.

Even after these economies, the money still would not stretch. That backlog of data Galton had been desperate to get his hands on was proving more problematical than he had envisaged. FitzRoy's system of collating into data books, which he had so roundly criticised in his report, was jettisoned in favour of a more modern card index. After only a small fraction of data had been laboriously transcribed, it was apparent this process was going nowhere. It took far too long, and even with only the 700 hundred cards the staff had managed to extract thus far, sorting them manually to configure, say, a set of barometer readings for a specified area proved well-nigh impossible. Galton was forced to retreat back to shelves of data books.

New observations began to pile into the office more quickly than they could be processed. As for the backlog, which, despite FitzRoy's curtailment, still amounted to over 2,000 ships' registers, it was calculated it would take in the region of 8,000 man-days to clear. The committee discussed the option of diverting staff from the daily weather reporting, but, despite the committee's declaration that this was the least important of its tasks, it was the public face of the operation and could not be cut. There was, therefore, only one conclusion to be made. Reluctant though he must have been, Galton was persuaded to agree that, for the present, it was better not to be too anxious to collect new information, as the Office's staff were finding it impossible to cope with the mountains of figures they had already got. This was exactly what FitzRoy had been saying, every year, for the past seven years.

So, when the Board of Trade began to demand the reinstatement of storm warnings, the Meteorological Office was already struggling to make ends meet with the staff and the money it had. There was no way out for them, though. It was obvious something had to be done, and indeed there was a deal of sympathy within the Meteorological Committee for storm warnings. Even Galton had previously expressed his support (in the report of the Committee of Enquiry). But Sabine's scientific dogma had to be respected and therefore the reply to the Board of Trade began by stating very clearly that the committee declined to 'prognosticate weather or to transmit what have been called "storm" warnings'. However, the committee was prepared to 'furnish, without any unnecessary delay, any telegraphic information which it may have received. In the case of telegraphic communications of this nature, half of the expenses is to be borne by the local authorities.'

This was not nearly good enough for the members of the Board of Trade. FitzRoy, they said, had provided poor fishing villages with free signals, and it insisted that the Meteorological Office should now do the same. Reluctantly, the committee agreed, seeing more money sliding away and another of the longed-for automatic recording machines slipping further into the future. Also, complained the Board of Trade, FitzRoy's signals had indicated the strength and direction of the coming gale. All right, said Scott and the committee, we will work on the design of some new, improved signals, but they baulked at paying for them. So did the Board of Trade, and instead handed over its stock of FitzRoy's old cones and drums.

Time rolled on, and nothing appeared to be happening. At the end of October, the committee received a letter sent to the Duke of Richmond by Sir James D.H. Elphinstone, Bart. He told them that the latest wrecks returns had revealed the 'appalling facts' of the numbers of shipwrecks. 'It will be much regretted if this winter is allowed to come on without some such system of warnings being renewed as that which was so successful under Admiral FitzRoy's management.' Would storm warnings be resumed in time for winter storms? They could be recommenced with very little delay, replied the committee, but they would consist simply of a signal to hoist a drum, to indicate the presence of a gale. The telegram issued would simply be a weather report, on the lines of 'Storm west of Penzance and south coast. Hoist Signal'.

If the seaman seeing the signal wanted to know more, then he would have to rush to the telegraph station to ask what the report had been – and then guess whether it would be coming his way. But to provide this service, including the free telegrams to needy fisheries, the committee wanted back that £3,000 that had been docked from its budget.

> With a view of collection and distributing such information, the Committee included a sum of 3000*l* in their estimate, and they are willing to communicate information . . . to an extent limited only by the sum placed at their disposal for the purpose.

It would have to do. It was better than nothing. The Board of Trade asked the committee to prepare a circular for issue, and the drums were distributed via the Coastguard. In some places the signal poles had disappeared (in Valencia, Ireland, for example, the pole had been appropriated for use in the local regatta) and the town had hastily to scout around for a replacement. But finally, on 28 December 1867, storm warnings – or rather 'telegraphic intelligence' of storms – were resumed.

There is a subtle difference between storm warnings and storm intelligence. FitzRoy had always recognised it, and, if they were honest, the members of the Meteorological Committee did, too, but it was probably still completely lost on Farrer and the Board of Trade. It was that storm 'intelligence', or knowing that a storm was happening, was as good as useless to coasts in the west of the country, the very place where storms hit most. However, the fact that a system of some sort was in operation placated the populace, pleased the Board of Trade and sat well with the principles of the Royal Society.

It did not, however, deliver. All it managed to do was to prove that FitzRoy's constant objections to a storm-intelligence system were well founded. Ports on the east coast could mostly be warned in time, but the system gave no benefit to outlying coasts of Ireland and the West of Scotland. This the committee shrugged off by stating that 'fortunately these shores are little frequented by shipping'. However, as a sop to storm-ravaged Galway and the shipwrecked mariners of the Western Isles, it promised these areas would eventually benefit, since ongoing studies meant it would soon be possible to recognise an approaching storm by its signs. This promise must have sounded sour in the wake of FitzRoy, who for four years had provided them with just such a system from which they had benefited admirably. The most striking success, and one that the committee was proud to point out, was in the warnings sent to Hamburg: out of 37 telegrams, 19 had been followed by a storm, 9 by stiff breezes and 6 by high winds – although in 3 cases, the storms had apparently come from the east and arrived before the telegram. While no doubt happy for the burghers of Hamburg, Liverpool seamen plying their 'little-frequented' routes must have been left wondering at the sense of a British storm-warning service that saved German lives while leaving them to fare as best they could at the mercy of the elements.

While storm warnings and weather reports were the public face of the new Meteorological Office's work, behind the scenes the staff were desperately trying to cope with the chaos they had inherited from the overstretched FitzRoy. Not only had returned marine logs been left to moulder; the instruments themselves were allowed to wander unsupervised. Barometers and thermometers recorded as issued to ships long decommissioned had to be traced and retrieved, while instruments belonging to the Admiralty were discovered in operation on merchant ships and in telegraph stations. Robert Scott personally undertook a tour of the country, finding barometers covered in dust, thermometers kept in roofs and wet bulbs permanently immersed in water, or coated in limescale. Wind speeds were guessed at by untrained clerks who did not know a Beaufort scale from a weighing scale.

With observers newly instructed, broken instruments replaced and others cleaned and recalibrated, the Office could rely on greater accuracy of reporting and confidently began to build theory from readings. In 1869, a couple of years after it began its work, it won an increase in the number of clerks, ostensibly to conduct an investigation into weather over the equatorial portion of the Atlantic Ocean. By 1871, daily weather maps, though not yet forecasts, were being issued, showing isobars, isotherms (lines connecting places of equal temperature), wind speed and direction, and so on. Storm warning accuracy showed that 349 gales were signalled, around 30 of them admittedly late, but only 25 were missed altogether. Nearly a half of the warnings were followed by gales of force 8 or more, about a fifth were force 7, while there was a good proportion, 22.4 per cent, that were not justified at all.

When compared to FitzRoy's accuracy, it is difficult to see how the Office could claim success for its system. Even based on Galton's figures quoted in the 1866 Report of the Committee of Enquiry, which were blatantly weighted against him, FitzRoy was achieving at least as good a record as the new Office. In fact, FitzRoy's legacy continued to form the backbone of the Met Office's public operations. Observations were still gathered by instruments based on his original design, fishing communities still benefited from their FitzRoy barometers and *Barometer Manual* (although this did receive updates with improved observational conclusions) and storms were still warned by the hoisting of his drum, to be joined once again by the directional cone in 1873. Captain Toynbee, who had designed a more complex system of semaphore arms to signal storm information, was forced to concede after a couple of years of trials that FitzRoy's robust, omni-directional signals could not be bettered.

Come 1872, the world was ready for another attempt at an international meteorological conference, and a gathering of interested meteorologists from around the world was held in Leipzig. After twenty years of progress, it was recognised, as Maury had always claimed, that the rivalries that separated land observations from maritime were futile and counterproductive, since weather covered the whole of the planet. In 1873, a full Congress was called at Vienna, the objective being an ambitious scheme to collect synchronised observations over the whole of the globe. As an indication of the way the world now regarded meteorology, delegates were not naval, but were representatives of their country's meteorological services. Yet perhaps the most significant outcome of this conference, the lack of which could be seen as the biggest failing of previous international efforts to coordinate meteorological investigations, was the setting-up of a permanent committee, initially headed by Buys Ballot and with Robert Scott as Secretary. At a later Congress in Rome, in 1879, this committee was to be ratified as the International Meteorological Committee, the directing committee of the new International Meteorological Organisation, which would lead world cooperation in meteorology well into the mid-twentieth century.

The American representative at the Vienna Congress was General Albert J. Myer, director of the US weather service. Myer was not a scientist; he was an army man, having begun his service as an assistant surgeon in 1854. He soon discovered that his interests lay more in the meteorological than the medical line, when he found himself taking a keen interest in the weather reports that it was his task to file. At the outbreak of Civil War, Myer was entrusted with organising the Signal Corps, cutting Joseph Henry off from his meteorological observers as the telegraph network was monopolised by military communications. After the war, the Signal Corps dwindled, leaving Myer in command of little more than a man and a boy, but he spied the saving of the corps when the establishment of a national weather service was called for by Joseph Henry and others. Campaigning on the basis that military signalling posts could track the attack of a storm just as effectively as they spied on enemy advances, he won the service for the Army Signal Corps and set up the first American Weather Bureau. As its director, Myer attended the world conference in Vienna.

Noticeably absent from Vienna were the French. Le Verrier and Marié-Davy had both received individual invitations, but both men were caught up in the tortuous politics of the French Imperial Observatory. In early 1870, Le Verrier's autocratic regime at the observatory had finally pushed the staff to rebellion, and he had been dismissed from office. Marié-Davy, meanwhile, was languishing at the Montsouris Observatory, an establishment only recently set up by Napoleon III as the equivalent of the Kew Observatory in England – a meteorological and magnetic institution, with no astronomical connections. Le Verrier's successor was keen to rid the Imperial Observatory of its meteorological work, leaving Marié-Davy as what would seem to be a strong contender to take the lead once more in French meteorology. Unfortunately for him, this was not to be.

The power struggles at the top of the meteorological establishment were as nothing compared to the tumult that was about to rage through the whole of France when, in the autumn of 1870, the country was plunged into war with Prussia. In December, the French Army was defeated at Sedan, Napoleon III abdicated, and Paris was laid to siege. Marié-Davy at Montsouris managed to escape to the provinces, from where he manfully succeeding in maintaining a creditable storm-warning service in the face of massive difficulties. After peace with Prussia, Marié-Davy became chief of the meteorological service at Montsouris, now an official outstation of the newly renamed Paris Observatory. His elevated status did not last long, destroyed by a boating accident in which the new director of the observatory was drowned. The way was open for Le Verrier to march in once more. In February 1873 he installed himself in his old position at the observatory and effectively sidelined Marié-Davy by grabbing back the national meteorological service and demoting Montsouris to a mere municipal establishment. It was little wonder that, in the midst of these internal upheavals, the matter of an invitation to Vienna fell by the wayside.

Robert Scott, on the other hand, as Secretary to the International Meteorological Committee, was busy taking his place as a world player in meteorology, while under his leadership the work of the Meteorological Office was progressing smoothly, certainly in comparison to the French. By 1873, 530 copies of the daily weather chart were being distributed, and when, two years later, the technology had been developed to allow graphics to be printed in newspapers, the first weather map appeared in *The Times*.

However, the accuracy of storm warnings was not improving, or at least not by much. In truth, the Met Office had probably gone about as far as it could go, hidebound as it was by the Royal Society's insistence on using only storm intelligence and not prediction. Each year since its inception, the Treasury had voted the Meteorological Committee of the Royal Society an amount of funding, and given it the freedom to spend the money as it wished, aside from the dictated provision of storm warnings. Come July 1875, the Treasury thought it time it reined in the power it had lost over its own expenditure. An inquiry into the Meteorological Office was decreed. The Report, issued in February 1877, recommended that the current arrangement with the Royal Society should cease, and its committee be replaced by a paid council. The

Meteorological Committee of the Royal Society resigned en masse in 1877.

On the face of it, little really changed. The new council had a chairman, one of its members was from the Admiralty Hydrographer's Office and the remaining four members were nominated by the Royal Society. Francis Galton transferred from the committee to the council as one of those members. Robert Scott became nominally Secretary to the new council, although his work and responsibilities remained exactly as they were. There was a subtle difference, though, in that, while the committee members, Scott apart, had been unsalaried volunteers operating under the auspices of the Royal Society, the members of the council were paid by, and therefore responsible to, the Treasury. Whether this lifting of Royal Society control gave them the freedom, or whether the reason was, as they declared, that the science of meteorology had progressed sufficiently to allow it, the result was the same. Just a year after its inception, the council felt sufficiently emboldened to announce that the Meteorological Office was ready to begin the reintroduction of weather forecasts.

In truth, its members had been practising their forecasting skills in private for a number of years. Robert Scott, head of the Meteorological Office, would have been familiar with the theories on which FitzRoy and Babington had based their forecasts, since they stemmed from Dové, whose work he had studied. While he might have been a little mystified by the lunisolar effect on which FitzRoy placed such store, he would have understood the principles of conflicting polar and equatorial air streams, and the way they were thought to clash in the temperate zones of the North Atlantic to create the disturbances that led to storms. Scott, though, was open to another influence. In the 1850s, Buys Ballot of Holland had intuitively constructed a rule for the relationship between barometric pressure, wind speed and direction, which had first been published in 1857 and came to be known as 'Buys Ballot's Law'. As explained by Buys Ballot to the British Association in 1863, his rule stated that, if you stand facing the wind, to your left will be low pressure and to your right, high, and that the greater the difference in pressure between two neighbouring points, the stronger the wind would blow. In other words, the wind does not blow directly down the incline from high to low pressure, but at right angles to it – a principle we see illustrated clearly on our weather charts today, where the wind direction arrows flow with the lines of isobars, rather than across them, and the wind is stronger where the isobars are closely packed. Based on his rules, Buys Ballot produced forecasts that he claimed to be on a par with FitzRoy's, despite having only four reporting stations in Holland as opposed to FitzRoy's twenty in Britain.

Although Buys Ballot speculated on the laws of nature governing his rules, it was an American, William Ferrel, who, quite independently, had first attempted the definition of a formula. Ferrel was a schoolteacher in Nashville, a self-taught mathematician with an interest in physical science. Marooned amidst the hills of Tennessee, he did not circulate within the coterie of the meteorological elite. Instead he educated himself, devouring the works of respected scientists, people such as Espy, and the man who had also spent his boyhood near Nashville, Matthew Maury. Ferrel, like many, saw little merit in Maury's fuzzy reasoning and it was partly Maury's faulty attempt at a theory of atmospheric circulation that prompted him to investigate the subject himself. In 1856, he wrote an essay – 'Essay on the Winds and Currents of the Ocean' – in which he identified a force hitherto ignored, a force created by the earth's rotation, which deflects a moving object to the right north of the equator and to the left in the south. Ferrel's ground-breaking discovery had the potential to revolutionise the theory of circulation at a time when men all over the world – Dové, Maury, FitzRoy, Buys Ballot and others – were grappling to make sense of the earth's winds, but unfortunately the work was published in the obscure *Nashville Journal of Medicine and Surgery*, and thus lay hidden to science. There is some evidence of news of Ferrel's ideas leaking across to Europe, but little attention appears to have been paid to them. FitzRoy certainly was aware of his work, although he probably had not understood it; and anyway, he was blinkered by his opposition to the use of isobars, partially because Galton approved of them but also for the valid reason that insufficient points of measurement rendered them unreliable.

Robert Scott, on the other hand, striving to find a basis for forecasting sufficiently scientific to win the approval of the Royal Society, latched onto Buys Ballot's rules and Ferrel's formulaic theories in preference to FitzRoy's more intuitive methods. For his research, he was helped by the ever-increasing number of returns coming into the Office, boosted in 1874 by the cooperation of the British Meteorological Society. For some reason, even though FitzRoy was nominally a member, the Meteorological Society had always stood at a discreet distance from the Meteorological Office. Historically, this would have been because of the Office's maritime function, but latterly had more to do with the society's disapproval in general of the Office's unscientific antics and the downright antagonism between Glaisher, the society's Secretary, and FitzRoy. Now that the whole operation was run on a footing of which it approved, the society began to provide the Office with registers of its observations, encouraged by the promise of free weather charts in return. The bulk of these were scattered and retrospective – amateur observers were rarely prepared to return their observations at set times, and certainly not by telegraph. For purposes of research, this did not matter, and they formed useful material for retrospective forecasting. Gradually the lessons learnt were put into practice and slowly Scott and his colleagues gained confidence in their ability to hazard forecasts at least as reliable as FitzRoy's – with the added advantage of a foundation of properly constructed maxims.

At last, on 1 April 1879, eighteen years after FitzRoy's original forecasts had appeared and almost thirteen years after they had been suspended, daily weather forecasts were made available to the British public once more. Drawn three times a day, they were supplied on subscription to anyone who wanted them, both private individuals and the press. Additionally, for a fee of three shillings, the public could telegraph or call in person at the office for their own personal twenty-four-hour forecasts. In the Meteorological Committee's Report of 1879, it was thought that the issue of a mere 418 of these individual requests proved that there was little demand for private information, and with forecasts appearing, for free, morning and evening in the newspapers, this was no doubt the case. Printed in newspapers under the daily weather map, with the outlook for various parts of Britain, the forecast took on, for the first time in history, the familiar shape and form that we identify today.

In 1879, forecasting the weather came of age. Not quite science, not quite art, more pragmatic than precise, it had travelled the road from myth to maturity. No longer was it castigated as comic, no better than a fairground fortune-teller and needing a crystal ball and bundle of seaweed to succeed. It had survived its initiation and, even if it was not yet their equal, it could stand alongside other sciences as an honourable and worthy pursuit. However, it still had far to go.

A New Dimension

The time is coming – and my plan will hasten it – when these
[weather] 'probabilities' will become certainties, and be more
specific and practical.

Matthew Maury's optimistic words were uttered in 1871, in
the climate of a country rebuilding itself after years of
internecine war. The first official American weather
forecasts, known then as weather 'probabilities' (and to
which Maury was referring), had just been issued and he was
beginning once more to push his plan for agricultural
meteorology, which had been shelved at the outbreak of war.
The Civil War had halted non-essential scientific work
everywhere. Even the Smithsonian Institution was not
immune, Joseph Henry being abruptly cut off from his
network of observers when the telegraph was wholly
appropriated for military use.

The end of hostilities did not mean an immediate
resumption of normal life. When Maury sailed from London
two days after FitzRoy's death, he did not make it back to
Virginia. Unknown to him, General Lee had surrendered the
week before he left England, and it was far too dangerous for
him, as a Confederate officer, to return home. He diverted to
Mexico instead. His intention was to establish a colony for
displaced Southerners in Montezuma, and he sent for his
wife and daughters to join him. They, however, despite being
nearly destitute, did not fancy a primitive, pioneering life
and refused to go. Therefore he begged them to leave for the
security of England, an idea they found much more to their
liking, and later he sailed to join them. In England, he lived
by writing geography textbooks (Maury's skills as a
geographer were renowned from his *Physical Geography of
the Sea*), augmented by the generous bounty of a 3,000

guinea memorial fund, which had been collected from his European admirers by the Revd Tremlett. Following the general amnesty of 1868, he returned to America, taking the Chair of Physics at Lexington, the Virginia Military Institute. Eventually, he regained sufficient academic stature to resume his plans for agricultural meteorology, and set off again on extensive tours of the Agricultural Societies. It was a strenuous undertaking for a man now well into his sixties. Taken ill after one of these lectures, he died in Lexington in February 1873. He was fortunate, perhaps, not to live to see that his hopes for turning 'probabilities' to 'certainties' were not to be realised.

After the war, when the national weather service was established under the Army Signal Corps, its director, Myer, cast around for scientific advice on the setting-up of regional signal stations. In Ohio he came upon the undoubtedly talented Professor Cleveland Abbe of the Cincinnati Observatory. By 1869 Abbe had already established his own network of telegraphed weather observations with the cooperation of the Western Union Telegraph Company and was issuing regular local weather reports and forecasts. Impressed, Myer recruited Abbe to the corps as civilian scientific adviser, from where he began to predict the weather. It is Cleveland Abbe who perhaps truly should be recognised as the father of American weather forecasting. His early 'probabilities' were intended to be warnings of storms to shipping on the Great Lakes and on the coasts. By 1876 'probabilities' had changed their name to 'indications' and were being issued generally to business and agriculture as well as to shipping. However, it was not until 1889 that Robert FitzRoy's term 'forecast' was officially adopted by the American service.

By 1890 forecasts were so widely issued and so many observations transmitted daily that the amount of telegraph traffic was thought a threat to the Army's signalling capabilities and might therefore jeopardise national security. As a result, in 1891 Abbe and his civilian scientific staff were transferred from the Army and formed the USA's first Weather Bureau under the auspices of the Department of Agriculture. Maury would have felt justly vindicated had he lived to see agriculture thus placed at the heart of the nation's weather services.

Cleveland Abbe's fame as a meteorologist spread worldwide as a result of his exhaustive support of the international simultaneous-observations exercise, the major initiative to emerge from the 1873 Congress of Vienna. Another vital, but much more technical, spur to progress was his translation of the mathematical relationship between wind speed and the spacing of isobars. The complex workings, and the formula derived from them, had first been deduced by a French civil engineer, Peslin, operating totally independently of meteorologists such as Le Verrier and Marié-Davy. Though it had been published in the *Bulletin international* in 1872, no one had taken particular account of Peslin's work until Abbe produced his explanation and understanding followed. Peslin's formula was eventually to form the basis of wind-speed calculation in modern dynamic meteorology.

Meanwhile, in France, Le Verrier's second reign at the Paris Observatory came to an end in 1877, when he died after a year of struggling against illness, an illness that must have caused inordinate suffering, too, among his poor staff forced to labour in the face of a vitriolic temper embittered by pain. With Le Verrier gone, the Ministry of Public Instruction consulted the Academy of Sciences on the wisdom of creating a totally separate meteorological establishment, as operated then by just about every other civilised country in the world. The academy was totally against the move – so the ministry ignored it, went its own way and appointed E.E.N. Mascart, scientist but non-meteorologist, to head the new Bureau Central Météorologique.

In the quarter-century since Maury's Brussels Conference, weather forecasting had become established as a recognised branch of meteorology. Through the International Meteorological Organisation, meteorological services all over the world were cooperating in observations, and producing daily storm warnings, reports and forecasts that improved the lives of everyone, from the sailor setting out to sea, to the farmer planting and harvesting his crops, and the traveller wishing to plan a journey. For the next quarter-century, although methods improved, observational networks grew and understanding increased, there was little in the way of major progress. Forecasting still relied on too few observations, entrenched in an inadequate two-dimensional view of what is in reality a multilayered atmosphere. And so it was set to stay, until the turn of the twentieth century.

The initial boost to meteorology in the new century, for Britain at least, was to come from an injection of science. The Meteorological Council, first appointed after the Treasury investigation in 1877, still ruled the Meteorological Office. Here, very little had changed in over twenty-three years. The Office continued as a more-or-less clerical operation, wading and sifting its way through thousands of observations; an unimaginative and increasingly narrow-minded Robert Scott remained as its director; and an ageing Francis Galton retained his position as senior member of a council still reluctant to embrace forecasting as a fully legitimate activity. Then, in 1900, Scott retired. Appointed in his place was William Napier Shaw. Shaw was a scientist, a lecturer in Experimental Physics at Cambridge, and he whistled into the Meteorological Office with the force of a hurricane. He brought trained scientists onto the staff and initiated new studies; he moved to a new site with room for proper library and research facilities; and in 1905, when the Meteorological Council was abolished, he assumed complete control of the Office.

Elsewhere, events were afoot that would transform not just meteorology, but the life of nations. In Italy, young Guglielmo Marconi had spent the final years of the nineteenth century attempting to send wireless messages between his house and a receiver in his garden. His experiment succeeded, yet he failed to raise backing from the Italians, so he packed his bags, his transmitter and his receiver and made the journey to England. There, the Postmaster General, Sir William Preece, took him under his wing and, on 11 December 1901, the three dots of Morse's letter 'S' crossed the Atlantic from Poldhu on Cornwall's Lizard peninsular to a kite-borne aerial flying above St Johns in Newfoundland. The first transatlantic wireless telegraph message had been transmitted.

Meanwhile, secreted away at their home in Ohio, American brothers, bicycle engineers Wilbur and Orville Wright, were working hard to perfect an incredible new machine. Two years later, just before Christmas 1903, they wheeled their strange-looking vehicle onto the beach at Kitty Hawk, North Carolina, and, with Orville perched at its controls, it raced along the sand and took to the air in man's first heavier-than-air powered flight. The aviation era had begun.

Whereas it had been the needs of those sailing the seas that had prompted the nineteenth-century development of forecasting, in the twentieth century it was to be the pilots of the skies who would spur meteorological progress. The aviators, quite literally, dragged meteorology into a new dimension. For them, it was not enough just to know the weather at the surface. They needed information on the weather above them – what were the prevailing winds in the upper air, how low and thick were the clouds, where were the squalls, was there any fog, would there be risk of icing? And as planes soared ever higher, even more was required of the forecaster. Meteorology rose to the challenge.

In order to forecast in this new dimension, meteorologists needed to learn more about it. Since the first weather flight of Dr John Jeffries and Monsieur Jean-Pierre Blanchard in 1784, balloons had provided grounded scientists with their best information on the state of the air above them. James Glaisher's 7-mile-high fright of a flight in 1862, though, was still man's deepest probe into the upper atmosphere. By the turn of the century, observers were sending self-recording meteorographs up into the sky attached to kites, tethered balloons, or to unmanned, free-flight 'pilot' balloons, which they had to trust to luck and the surrounding populace to retrieve. Compared with the daily, organised, surface observations now pouring into centres worldwide, upper-air data was at best spasmodic.

The First World War, coming just over ten years after that first shaky lift-off of Orville Wright's, dragged warfare into the air for the first time in history. With it, meteorology entered the arena of war as a strategic defence mechanism. In 1915 Napier Shaw released two of his Meteorological Office staff to go to France to form the nucleus of the new Meteorological Unit of the Royal Engineers. They were to provide meteorological reports and forecasts for military offensives on the ground and in the air. One of these men was Ernest Gold. Gold had already cemented his position as a meteorologist of note by progressing the work of Buys Ballot, Peslin and Cleveland Abbe in the calculation of wind speed from air-pressure differential, work that he further built upon in the field, accurately to predict winds for air navigators.

The war concentrated meteorological effort, established it on a true scientific basis through the efforts of men like Napier Shaw and Gold and for the first time provided sound, three-dimensional data. At its cessation, it was obvious the

Meteorological Office was ripe for revamping. After the vital role played in the war by aviators and the importance of weather forecasting to flying, Winston Churchill, Minister for Air and War, insisted that the Office should lose its independent status and be brought under the control of the Air Force Department. Napier Shaw fought hard to resist this move, but when it became inevitable, he announced his retirement, leaving his post in April 1920. The end of the war released planes and pilots, men such as ex-Royal Flying Corps captain C.K.M. Douglas, ready and eager to further the study of meteorology in peacetime. Now, with the aeronautical element firmly established as a priority, Ernest Gold created the specialist Meteorological Flight, and Charles Douglas became a member. The focus of their study was to identify the movements of air masses and the source of upper-air winds, both of which appeared to influence changes in the weather. They were not alone in their studies. Across the North Sea, in Bergen, a group of Norwegian scientists had been investigating the same phenomena and in 1920 they invited Napier Shaw (shortly before his retirement), along with Gold, Douglas and others, to visit Bergen to discuss their revolutionary new theories on the dynamics of the atmosphere.

Vilhelm Bjerknes had organised a system of weather observation stations throughout Norway during the war and, working with his son, Jakob, and other colleagues at the University of Bergen, he had investigated what the readings were telling them about the upper air. They discovered that a band of rain and cloud marked a boundary where opposing masses of air vied with each other, one warm, the other cold, and they gave the term 'polar front' to the leading edge of the advancing mass of cold air. The cold mass moved faster than the warm, as if chasing it, and pushed under the warm air, creating the distinctive sloping nature of the front. The warm air, forced to rise above the cold, shed its moisture in cloud and precipitation. Bjerknes's discoveries were to launch a whole new raft of theories that would come to be known as the Bergen, or Norwegian, School. These theories, applied to the two-dimensional weather maps of the day, brought with them a huge advance in forecasting, but they would show their full potential only when technology was sufficiently advanced to allow the construction and manipulation of three-dimensional models of the atmosphere.

One of the visiting party with Shaw and Douglas at Bergen was meteorological mathematician L.F. Richardson. Inspired by what he had seen, he put forward the idea that the entire atmosphere could be modelled by a complex set of mathematical equations, and that by adding a fourth dimension, time, the model could be used to extrapolate the movement of the atmosphere over a set future period. This was exactly the mode of mathematical forecasting that would have filled Francis Galton with joy. Richardson set teams of computers to work on his calculations. Unfortunately, in 1922, a 'computer' was simply a human being with pencil and paper. They took days to complete a twenty-four-hour forecast, finishing long after the weather had been and gone – and even then, their forecast was wrong. Still, the possibility was proved and a precedent set. Richardson dreamed of a 'forecast factory', where thousands of human computers would sit in a huge, globe-shaped room, each calculating the future weather in one small portion of the world. His vision would one day be achieved, but by electronic rather than by human means.

Not everyone subscribed to Bjerknes's three-dimensional ideas, preferring their tried-and-tested methods. In the United States, the Weather Bureau was proving resistant to the Bergen School ideas, and, while the bureau was serving its traditional customers well, the new breed of aviators needed more accurate forecasts of conditions above ground. In 1918, the first regular US air mail service was established, but after a disastrous year in 1920–1 when nearly ninety aircraft came to grief and many pilots were killed, an ex-flying officer, Francis Reichelderfer, decided it was time to do something to help. Reichelderfer had been a naval pilot in the First World War. One day, off on a bombing mission, he had flown straight into the face of an unforeseen squall. After the fright of his life, he landed safely, resolved that, should he survive the conflict, he would make it his aim to do his utmost to improve aviation forecasts. As Director of the Navy's Aerological Service after the war, he met Carl Rossby, a forecaster with the Weather Bureau, who had studied at Bergen. Enthused by Rossby's views on upper-air forecasting, he approached the Daniel Guggenheimer Fund for the Promotion of Aeronautics, with a suggestion that a committee, separate from the overburdened Weather Bureau, be set up specifically to further aeronautical meteorology. Rossby was made chairman of the subsequently formed committee, with Reichelderfer a prominent member. Reichelderfer further developed and explored his interest in meteorology over the next few years, until, in 1938, he was appointed chief of the US Weather Bureau.

Bjerknes's study of the atmosphere had stemmed from his investigation into the effect of atmospheric conditions on the transmission of electrical waves, a problem that had captured his imagination when Marconi's early attempts at wireless telegraphy had been beset by atmospheric interference. As wireless technology improved, the International Meteorological Committee was not slow to grasp its implications. Ship-to-shore communication offered the answer to a prayer – the chance to fill that yawning void on the daily synoptic charts that was the Atlantic Ocean. The war interrupted transmissions from ships, but immediately afterwards international cooperation began again. The big liners thronging the busy transoceanic routes were enlisted to provide the bulk of ocean wireless meteorological reports. But it was not just at sea where radio proved its worth. No longer restricted by the positioning of cables and telegraph stations, observers were able to transmit their readings from all kinds of unpopulated areas – inhospitable mountains, parched deserts, remote forests and freezing snowfields. Wireless also solved the problem of obtaining regular upper-air readings. It was difficult and slow to retrieve readings from kites and balloons; planes could only go so high and were expensive. Then, from France, came the invention of the radiosonde. This was a small, disposable radio device that could be sent up into the atmosphere to transmit its readings back to a ground receiver, without having to be retrieved. By the late 1930s, networks for upper-air radio sounding had been established throughout the world.

With radio, aircraft and rapidly developing theory, forecasting was poised on the brink of a revolution. However, while countries seemed happy to coexist and cooperate in the exchange of weather data, their governments were set on confrontation. The onset of the Second World War plunged forecasters into the midst of the conflict, with a whole new set of problems to face. The most serious of these was the sudden interruption of the flow of observational information.

Radio is an open communication medium, fine in peacetime, but a security liability during war. Meteorological intelligence had become a vital weapon, and the only way to prevent such sensitive information falling into the hands of the enemy was either to stop it or to encode it. Suddenly, British weather charts were holed with blank patches over Germany, Austria, Finland, Russia – and, most vitally, the Atlantic Ocean. To fill them, the Meteorological Office became a secret service, linked to Station X at Bletchley Park, and equipped and manned with its own staff of code-breakers. Beating the enemy to meteorological information became a means to defeating them in battle. Charles Douglas, the First World War fighter pilot turned meteorologist, and now senior forecaster, was to prove just how important this role was when he managed to predict the break in the weather – not spotted by his German counterpart – that allowed the optimum timing for the D-Day landings.

War, yet again, proved to be the spur for the next wave of developments in meteorology. Pilots, flying high into the atmosphere, found themselves buffeted by high-velocity winds never encountered before. They had discovered the jet streams, bands of strong winds, thousands of miles long and blowing up to 300 miles per hour. One such, the polar jet stream, flows at around 35,000 feet from west to east in the northern temperate zones, separating the cold polar air from the warm tropical air. This is the wind that propels jet airliners across the Atlantic, cutting flying times on the America-to-Europe trip, and elongating the journey back. Gradually, it was realised that depressions, which, it had always been believed, developed from conditions on the surface, could actually be formed from above by activity in the jet stream, and the effect on forecasting was as profound as Bjerknes's discovery of polar fronts.

Bletchley Park, with its demand for number crunching and through the inspired pioneering work of Alan Turing, gave birth to Colossus, the machine that would eventually evolve into the electronic computer. By the end of the 1950s, the first commercial computers began to give to meteorology the calculating power it had long hungered for – in 1959 the Meteorological Office in Britain installed its Ferranti Mercury machine, named *Meteor*. Jet and rocket technology, invented for offensive purposes during the war, was diverted and developed in peacetime to power the space race. In 1960 the first weather satellite orbited the earth, FitzRoy's 'eye in the sky' come into being, sending back pictures from space.

Today, forecasting is an intensely mathematical process. The complex interactions of the atmosphere are represented by thousands of equations, built into numerical models, such as the Meteorological Office's new Numerical Weather Prediction (NWP) model. The model's horizontal geographical base is one that both Maury and FitzRoy would recognise, dividing the globe into grid squares. However, each square is also divided vertically into levels, and as computing power expands, the model is further defined and refined into smaller grids and increased numbers of levels. Numeric models feed on millions of observational readings every day. The majority of these still come from minimalist land weather 'stations' returning measurements of temperature, barometric pressure, rainfall and wind speed, much the same as in FitzRoy's day. Marine observations are provided by ships and automatic buoys, while the upper air is measured by specially equipped aircraft and by radiosondes carried high into the atmosphere by helium balloons. Way above the earth, satellites measure winds and atmospheric motion, track vapour as clouds, rain or snow, assess temperature at all depths, penetrate the sea and ice, and generally fill in the gaps left in the data model from the uneven coverage of other sources. Yet, despite copious data and complex modelling, the final result still relies on the skill of a human forecaster, who notes current conditions and gives a personal interpretation of the computer's estimation.

It is difficult, now, to imagine a world without the weather forecast. Instantly available, it can be delivered by television, radio, newspaper, Internet, telephone or text message to a mobile phone. Multifaceted, a forecast can be tailored to meet specialist needs – aviation forecasts for aircraft of any size, from microlights to mighty jumbos; shipping forecasts for ocean-going craft and inshore forecasts for yachtsmen and fishermen; mountain area forecasts for climbers and walkers; personalised one-to-one forecasts through 'talk-to-the-forecaster' schemes; long-range forecasts tailored to the needs of agriculture, farming and industry; warnings of floods and storms; forecasts to prepare ice-cream manufacturers for a coming hot spell, or local-authority road-gritters for impending cold; and historical services, such as the Met Office's 'GeoProof' scheme, to provide the insurance industry with details of past weather events in order to assess weather-dependent claims.

And what about the future? As ever-more powerful computers crunch their way through increasingly numerous and complex equations, that search for precision in forecasting that possessed FitzRoy, Abbe, Napier Shaw, Bjerknes, Douglas and others still spurs on an army of mathematicians, physicists and forecasters worldwide. Developments of the past century and a half have brought about a phenomenal increase in speed and sophistication compared to FitzRoy's primitive methods. So why has accuracy not improved by the same degree?

The problem with the weather is that the only thing consistent about it is its unpredictability. The complex relationship between its major factors – pressure, temperature, humidity and wind speed – offers an inestimable number of different permutations. Early in the 1960s Edward Lorenz discovered that a minute change in just one factor of his computerised model of the global weather system produced immense differences in the results – the first identifiable instance of the 'chaotic' behaviour inherent in the model, behaviour that prompted his proposition 'Does the flap of a butterfly's wings in Brazil set off a tornado in Texas?' Chaos Theory or no, it is true that, even with today's complex computer models, it is not possible to project further than a few days into the future with any degree of accuracy, and even short-range predictions can prove wrong when the weather is affected by factors that were missed. There are those rogue days, still, when country lore will triumph.

The route to the forecast, which began with the Stone Age hunter, was paved by men, ideas and developments each as vital as the other. The milestones along the way were not all scientific, and in fact it is probably true that the art of forecasting, as opposed to the science of meteorology, has been spurred on more by humanitarian considerations than by the desire for scientific breakthrough. Galileo's thermometer and Torricelli's barometer provided the tools with which to work; it was Robert Boyle's discoveries with gases and Isaac Newton's laws of motion that set the movements of the atmosphere free from the restraints of Aristotle; the theories of Dové, Buys Ballot and Ferrel, and the intense debates between Espy, Henry and Redfield, all boosted the reputation of meteorology and gave it scientific respectability; but it was left to the practical men, those such as Maury and FitzRoy, to take the tools in hand, veer from the road of science and head off in aid of their fellow man.

As for the forecast itself, Admiral Robert FitzRoy is indubitably the man who invented it, since it was he who first coined the term 'forecast'. FitzRoy did not set out to instigate a daily forecast, but, from the day he took up the reins of the Meteorological Office, it was inevitable that would be the direction he would take. His determination to champion the humble sailor drove him towards a storm-warning service just as surely as his ship was driven by the wind. It also drove him to his own destruction.

It is fitting, therefore, to allow Admiral FitzRoy the final words. On Friday, 11 April 1862, less than a year after the publication of the admiral's very first forecast, this passage appeared in *The Times* leader, neatly summing up the origins of weather forecasting.

It is not necessary to impress upon a people whose earliest salutation in the morning and whose common topic of talk throughout the day is derived from the Weather and its operation upon all their avocations, how useful it would be if we could get a few hours' notice of its probable vicissitudes. The desire is so general that it is everywhere the parent of a hundred superstitions. Some men swear faithfully by a weather-glass which has gone on deceiving them for 40 years; others, more cynical, consult their barometer, and believe the opposite . . . Now, however, an attempt is being made to treat this matter in a scientific way, and to see whether we really cannot arrive at some approximate knowledge of the laws by which the fall of rain is regulated. In so novel an inquiry, of course our first guesses must be but guesses. All that is required at setting out is that our observations should be made with care and impartiality, and that our forecasts of the future should be made upon some principle which the results may either establish or destroy.

'Certain it is,' says Admiral FitzRoy, 'that, although our conclusions may be incorrect, our judgement erroneous, the laws of nature and the signs afforded to man are invariably true. Accurate interpretation is the real deficiency.'

We are still striving for that 'accurate interpretation' today.

Notes

METEOROLOGICAL OFFICE REPORTS

The Meteorological Office reports quoted in this book are held in the archives of the National Meteorological Library and Archive, at the Meteorological Office, Exeter. Apart from FitzRoy's unpublished report of 1855 and the Appendix to the 1862 report, copies are also held within the series of microfilmed Parliamentary Papers at the National Archives at Kew. The National Archives references are:

1857 Report	PP (1857) XX.238 Fiche No. 62.160
1858 Report	PP (1857–58) XXIV.389 Fiche No. 63.216
1862 Report	PP (1852) LIV.433 Fiche Nos 68.412–413
1863 Report	PP(1863) LXIII.27 Fiche Nos 69.493–494
1864 Report	PP(1864) LV.125 Fiche Nos 70.449–450
1867 Report	PP(1867) LXIII.295 Fiche Nos 74.723–724
1868 Report	PP(1868) XXIII.57 Fiche Nos 75.179–180
1869 Report	PP(1870) XXVII.17 Fiche No. 76.232
1870 Report	PP(1871) XXIV.17 Fiche No. 77.207
1871 Report	PP(1872) XXIV.145 Fiche Nos 78.203–204
1872 Report	PP(1873) XXIV.383 Fiche Nos 79.213–214
1873 Report	PP(1874) XXI.15 Fiche No. 80.156
1874 Report	PP(1875) XXVII.17 Fiche No. 81.200–201
1875 Report	PP(1876) XXVI.15 Fiche No. 82.174
1876–77 Report	PP(1877) XXXIII.631 Fiche No. 83.244–245

ABBREVIATIONS

The following abbreviations are used within these notes:

NMLA	National Meteorological Library and Archive, Exeter
PP	Parliamentary Papers
TNA	The National Archives (formerly Public Record Office), Kew

PROLOGUE

[1]   'It began at seven in the morning . . .': Admiral Heath, in a letter written on board HMS *Sanspareil*, 23 November 1854, nine days after the Black Sea storm; in Admiral Sir Leopold George Heath, KCB, *Letters from the Black Sea during the Crimean War 1854–55* (London, Bentley, 1897), p. 109.

[1]   'During the squalls . . .': Revd S. Kelson Stothert to his mother, from HMS *Queen* off the Katcha, 17 November 1854; in Mrs Tom Kelly, *From the Fleet in the Fifties: Incorporating Letters by the Reverend S. Kelson Stothert, Chaplain to the Naval Brigade* (London, Hurst & Blackett Ltd, 1902), p. 260.

[2]   'a most beautiful little harbour . . .': Captain Nicholas Dunscombe, 46th (South Devonshire) Foot, Sebastopol, Crimea, Wednesday, 8 November 1854; in Major Colin Robins (ed.), *Captain Dunscombe's Diary, The Diary of Captain Nicholas Dunscombe, 46th (South Devonshire) Foot* (Bourdon, Withycut House, 2003), p. 41.

[2]   'It rained, hailed and snowed . . .': Captain Nicholas Dunscombe, Sebastopol, Crimea, Tuesday, 14 November 1854; in *ibid.*, pp. 43–4.

[3]   'If this weather continues . . .': Captain Nicholas Dunscombe, Sebastopol, Crimea, Tuesday, 14 November 1854; in *ibid.*, p. 44.

[3]   'I could not help reflecting . . .': Revd S. Kelson Stothert to his mother, from HMS *Queen* off the Katcha, 17 November 1854; in Mrs Tom Kelly, *From the Fleet in the Fifties*, p. 260.

CHAPTER 1. MYTHS TO METEOROLOGY

[12]   'An ass shaking its ear . . .': This and the following quotations are taken from *Theothrastus of Eresus on Winds and on Weather Signs*, trans. J.G. Wood, MA, LL.B., FGS (Stanford, 1894); in Sir W. Napier Shaw, *Manual of Meteorology*, vol. i (Cambridge, Cambridge University Press, 1926), pp. 99–100.

CHAPTER 2. WEATHER WATCHERS

[20]   'The summer of 1783 was an amazing . . .': Gilbert White to Barrington no 65; in Gilbert White, *Natural History of Selbourne*, ed. with an introduction by Paul Foster (Oxford, Oxford University Press, 1998), p. 247.

[21] 'My pretensions as a man of science . . .': Luke Howard to Goethe, 1822; in D.F.S. Scott, *Luke Howard: His Correspondence with Goethe and his Continental Journey of 1816* (York, Williams Sessions, 1976), pp. 2–3.

[21] 'The numerous Aurorae boreales . . .': *ibid.*

[21] 'in passing between the works . . .': *ibid.*

[23] '[Clouds] are commonly as good . . .': Luke Howard, *Essay on the Modification of Clouds* (1st edn, 1804; repr. London, 1830).

[23] 'My friend Allen and myself . . .': Luke Howard to Goethe, 1822; in Scott, *Luke Howard*, pp. 2–3.

[23] 'in order to enable the Meteorologist . . .': Howard, *Essay on the Modifications of Clouds*.

[24] 'the facts needed for this art . . .': note written by Lavoisier, in Shaw, *Manual of Meteorology*, vol. i, p. 302, trans. P. Halford.

[27] 'This may in time either facilitate . . .': William Borlase to Henry Baker, 18 July 1863; in Vladimir Jankovic, *Reading the Skies: A Cultural History of English Weather 1650–1820* (Manchester: Manchester University Press, 2000), pp. 111–12. Baker was a microscopist, who advised Borlase on the use of meteorological instruments.

[30] 'Every thing, in time, becomes to him . . .': John Claridge, Introduction, in *The Shepherd of Banbury's Rules* (1670); in Shaw, *Manual of Meteorology*, vol. i, p. 109.

[30] '*if the sun rise red and fiery . . .*': This and the following quotations are taken from *ibid.*

CHAPTER 3. STORM ABOUT STORMS

[40] 'I hear from England that . . .': Reid to Redfield, 7 November 1840, Reid's correspondence, Yale University; in James Rodger Fleming, *Meteorology in Ameica 1800–1870* (Baltimore, Johns Hopkins University Press, 1980), p. 60.

CHAPTER 4. CHARTING THE WINDS

[51] 'To enable it to bring this undertaking . . .': Maury's appeal to the US National Institute for the Advancement of Science, 1843; in Chester G. Hearn, *Tracks in the Sea: Matthew Fontaine Maury and the Mapping of the Oceans* (London, International Marine/McGraw-Hill, 2002), p. 101.

CHAPTER 5. NATIONS CONFER

[54] 'You will please . . . give your views . . .': Commodore C. Morris, US Navy, to Lieutenant Maury, November 1851, PP (1852–53) LX, pp. 441–63.

[54] 'it would create confusion among . . .': Lieutenant Maury to Commodore Morris, 21 November 1851, PP (1852–53) LX, pp. 441–63.

[55] 'Five-sevenths of the surface . . .': *ibid.*

[55] 'The atmosphere envelopes the earth . . .': *ibid.*

[56] 'I beg leave to suggest a meteorological conference . . .': *ibid.*

[57] 'I would beg to submit . . .': Sir John Burgoyne to Mr Addington, Foreign Ofice, 25 February 1852, PP (1852–53) LX, pp. 441–63.

[58] 'correct climatological knowledge . . .': Lieutenant-Colonel Edward Sabine to Mr Addington, Foreign Office, 12 May 1852, PP (1852–53) LX, pp. 441–63.

[59] 'With reference to the proposal . . .': *ibid.*

[59] 'The proposition of Lieutenant Maury . . .': *ibid.*

[64] the very house, once called Whiteladies: Maury's daughter Diana reports on this, and the excitement of the trip to Europe in her book, Diana Fontaine Maury Corbin, *A Life of Matthew Fontaine Maury* (London, Sampson Low, Marston, Searle & Rivington, 1888), p. 156.

[65] 'The object of our meeting . . .': Minutes of the Sittings, Abstract of Report of Conference held at Brussels Respecting Meteorological Observations, PP (1854) XLII, pp. 443–74.

[66] 'The Science of Meteorology can now be successfully cultivated': Captain Henry James to Major-General Sir John Burgoyne, 14 September 1853; TNA BJ 7/106.

[66] 'The Conference, having brought to a close . . .' : Abstract of Report of Conference held at Brussels Respecting Meteorological Observations, PP (1854) XLII, pp. 443–74.

CHAPTER 6. WHO WILL BE THE WEATHERMAN?

[70] 'I was thus led to study the colours of the sky . . .': James Glaisher, Camille Flammarion, W. de Fonvielle and Gaston Tissandier, *Travels in the Air*, ed. James Glaisher (London, Richard Bentley, 1871), p. 29.

[70] 'the cause of their rapid formation . . .': *ibid.*, p. 29.

[72] 'In the whole world . . .': George Airy to Henry James, 7 October 1853; TNA BJ 7/108.

[72]   'I have no objection to undertaking it': George Airy to Henry
        James, 13 October 1853; TNA BJ 7/109.

CHAPTER 7. SURVEYING THE SEAS

[76]   'I never in my life met a man . . .': Charles Darwin to his family
        at home, from Bahia, Brazil, 8 February 1832; in Francis
        Darwin (ed.), *Life and Letters of Charles Darwin*, 4 vols
        (London, John Murray, 1887), vol. i, pp. 202–3.
[76]   'has been examined . . .': examination certificate from the Royal
        Naval College, Portsmouth, awarded 8 October 1817 to John
        Jarvis Tucker; National Maritime Museum, Greenwich, TUC/47.
[82]   'may be of use – but only if steadily and accurately kept':
        memorandum   compiled   by   Captain   Beaufort,
        accompanying the Admiralty Instructions issued for the
        second surveying voyage of the *Beagle*, 1831; National
        Maritime Museum, Greenwich.
[83]   'Their Lordships do not approve . . .' : Captain's papers,
        Admiralty archives; TNA ADM/1/1819.
[83]   'My schooner is *sold* . . .': Robert FitzRoy to Sir Francis Beaufort,
        26 September 1834; in David Stanbury (ed.), *A Narrative of the
        Voyage of HMS Beagle* (London, Folio Society, 1977), p. 209.
[83]   'the selling [of] the schooner . . .': Charles Darwin to his sister
        Catherine, from Valparaiso, 8 November 1834; in Francis Darwin
        (ed.), *Life and Letters of Charles Darwin*, vol. i, pp. 228–9.
[83]   'He has already regained his cool, inflexible manner . . .': *ibid.*
[86]   'neither had nor cared . . .': Robert FitzRoy, *The Weather Book:
        A Manual of Practical Meteorology* (London, Longman, Green,
        Longman, Roberts & Green, 1863), pp. 334–5.
[86]   'But about two o'clock . . .': *ibid.*

CHAPTER 8. THE BIRTH OF THE MET OFFICE

[88]   'zealous, trustworthy, efficient officer . . .': Robert FitzRoy,
        draft memorandum, 5 November 1853; TNA BJ 7/113.
[89]   'If I do not see by the papers . . .': Captain H. James to Captain
        F.W. Beechey, December 1853, TNA BJ 1/120.
[90]   'It would not be too difficult to carry on without a
        draughtsman . . .': Robert FitzRoy to Lord Wrottesley,
        President of the Royal Society, February 1854; TNA BJ 7/2.
[90]   'whether it was probable that an office . . .': Parl. Debs., 6
        February 1854, 3rd series, 130, col. 270, 'Improvements in
        Navigation – Captain Maury's plan'.

[90] 'in consequence of Captain Beechey's report . . .': *ibid.*
[94] 'The great extent of country . . .': James D. Forbes, 'Report upon the Recent Progress and Present State of Meteorology', *British Association Second Report* (1832), p. 235.
[96] 'that essential condition . . .': *ibid.*, p. 245.
[97] 'It is one of the chief points . . .': Robert FitzRoy, Report of the Meteorological Office, 1855; NMLA.
[99] 'anticipates much from FitzRoy . . .': Maury to Sabine, March 1854; TNA BJ 1/125.

CHAPTER 9. SIMULTANEOUS OBSERVATIONS

[101] 'it becomes my duty to inform you . . .': J.C. Dobbin to M.F. Maury, 17 September 1855; in Corbin, *A Life of Matthew Fontaine Maury*, p. 109.
[101] 'I have been brought into official disgrace . . .': M.F. Maury to the Right Reverend James H. Otey, DD, Bishop of Tennessee, 20 September 1855; in *ibid.*, pp. 111–18.
[102] 'I have great confidence in Maury . . .': Sir John Herschel to Robert FitzRoy, 10 April 1858; Herschel Papers, Royal Society Library.
[106] 'your paper on the motion of waves . . .': Robert FitzRoy to Henry Clifton Sorby of Sheffield, 26 June 1856; TNA BJ 7/549.
[109] 'The atmosphere is a great basin . . .': Maury's address to the US Agricultural Society in Washington, 10 January 1856; in Jaquelin Ambler Caskie, *Life and Letters of Matthew Fontaine Maury* (Richmond, Va., Richmond Press, 1928), pp. 77–9.
[112] 'Maury is now working out . . .': Robert FitzRoy to Robert Lowe, 9 January 1857; TNA BJ 7/553.
[113] 'I am enabled to write . . .': Robert FitzRoy, Report of the Meteorological Department of the Board of Trade, 1857; NMLA.
[113] 'simultaneous states of the atmosphere': *ibid.*
[115] 'Take notice now, that this plan of crop and weather reports . . .': Maury's address to the North Alabama Mechanical and Agricultural Society at Decatur, Alabama, 1856; in Caskie, *Life and Letters of Matthew Fontaine Maury*, pp. 75–6.

CHAPTER 10. THE FITZROY BAROMETER

[118] 'It is not from the point . . .': Robert FitzRoy, *Barometer and Weather Guide* (Board of Trade, 1858; 2nd edn., London, Eyre & Spottiswoode, 1859), p. 8.

[119] 'Long Foretold, long past . . .': *ibid.*, p. 19.

[119] 'When rise begins, after low . . .': *ibid.*

[120] Legend from FitzRoy barometer: FitzRoy, *The Weather Book*, p. 10.

[120] 'A Capital little Weather Guide': Matthew Maury to Robert FitzRoy, 1 January 1859; TNA BJ 7/95.

[122] Coastguards labelled all fishermen: Robert FitzRoy to H.R. Williams, Board of Trade Accountant, 6 August 1858; TNA BJ 7/17.

[123] 'give a succession of synoptic views . . .': Robert FitzRoy to T.H. Babington, March 1858; TNA BJ 7/564.

[123] 'we may elicit the nature and character . . .': *ibid.*

[123] 'as if an eye in space . . .': FitzRoy, *The Weather Book*, p. 103.

CHAPTER 11. SUN AND WIND

[125] 'Of all the problems in Meteorology . . .': J.D. Forbes, 'Report on Meteorology', Second Report of the British Association for the Advancement of Science, 1832, p. 236.

[127] 'what labours and dangers you have gone through . . .': Charles Darwin to Francis Galton, from Eastbourne, 24 July 1853, in Karl Pearson, *Life, Letters and Labours of Francis Galton* (Cambridge, Cambridge University Press, 1914), vol. i, p. 240.

[131] 'It is very desirable . . .': Lieutenant-Colonel William Reid, 'A Statement of the Progress made towards developing the Law of Storms', Eighth Report of the British Association for the Advancement of Science,1838, pp. 21–5.

[133] 'You will I am sure be gratified to see . . .': Maury to FitzRoy, February 1859; TNA BJ 7/96.

[134] 'In connexion with the setting up of Anemometers . . .': Robert FitzRoy, memo distributed from the Meteorological Office, 28–30 June 1859; TNA BJ 7/25.

[135] 'to institute a series . . .': *ibid.*

[135] 'the great collection of observations . . .': Robert FitzRoy, *Notes on Meteorology*, Board of Trade Miscellaneous Papers, London, HMSO, 1859; NMLA.

CHAPTER 12. ATMOSPHERIC GYRATION

[138] 'on the authority of Herschel and Dové': FitzRoy, *Notes on Meteorology*.

[139] 'that mysterious and subtle agent . . .': *ibid.*

[141] 'peculiar circumstance . . .': *ibid.*

[142] 'Among the objects to which meteorologists . . .': *ibid.*

[142] 'you will scarcely be mistaken . . .': *ibid.*

[142] 'Soon after a few of the earlier synoptic charts . . .': FitzRoy, *The Weather Book*, p. 107.

[143] The proposition had first been put forward in 1856 by American William Ferrel: in the *Nashville Journal of Medicine and Surgery*, and further developed in 1858 and 1859, see page 244; see also Fleming, *Meteorology in America*, pp. 138–9.

CHAPTER 13. *ROYAL CHARTER* STORM

[149] 'Having had many threatenings . . .': Robert FitzRoy, 'On British Storms', Report of the British Association for the Advancement of Science, 1860, p. 41.

[149] 'The simple rule of seamanship . . .': *ibid.*, p. 42.

[149] 'a ship managed . . .': *ibid.*

CHAPTER 14. STORM PREDICTION

[154] 'It is impossible for those who have studied practical meteorology . . .': Robert FitzRoy, Report of the Meteorological Department of the Board of Trade, 1862; NMLA.

[157] 'since meteorology is surrounded with uncertainties . . .': Robert FitzRoy to Milner Gibson, President of the Board of Trade, memo, 18 April 1860; TNA BJ 9/7.

[158] 'Si donc vous me le permettez . . .': Urbain Le Verrier to Robert FitzRoy, 18 April 1860, Appendix to the Report of the Meteorological Department of the Board of Trade, 1862; NMLA.

[158] 'seems to deprecate . . .': Thomas Farrer to Robert FitzRoy, 20 April 1860; in *ibid.*

[160] 'In this office . . .': Robert FitzRoy to Thomas Farrer and James Booth, memo, 24 April 1860, in *ibid.*

[160] 'a man might as well register . . .': Sir John Herschel, in *ibid.*

[160] 'and such an approximation . . .': *ibid.*

[161] FitzRoy . . . composed a straightforward précis: Robert FitzRoy to Milner Gibson, 30 April 1860, memo; in *ibid.*

[161] The President proclaimed: . . . Milner Gibson, memo, 6 June 1860; in *ibid.*

[162] 'with particular reference to Mr Darwin's work . . .': The two papers presented at the British Association meeting in Oxford, 1860, that precipitated the heated discussions on Darwin's theories were: Dr Daubeny of Oxford, 'On the Final Causes of

the Sexuality of Plants with Particular Reference to Mr Darwin's Work on the *Origin of Species*', Thursday, 28 June; and Dr Draper of New York, 'Intellectual Development of Europe Considered with Reference to the Views of Mr Darwin', Saturday, 30 June.

[162] 'The British Association has made application . . .' : FitzRoy, 'On British Storms', pp. 39–44.

[164] 'regretted the publication . . .': *Athenaeum* report of the 30 June debate at the Oxford meeting of the British Association, 1860; in H.E.L. Mellersh, *FitzRoy of the Beagle* (London, Rupert Hart-Davis, 1968), p. 274.

[164] 'I think his mind is often on the verge of insanity': Darwin to J.S. Henslow, 16 July 1860, Cambridge University Library, Darwin Papers 93, A74–75.

CHAPTER 15. FORECAST

[167] 'A column of your paper is filled today . . .': Robert FitzRoy, letter to *The Times*, 12 February 1861.

[167] 'I would repeat what has been reiterated . . .': *ibid.*

[168] The Lords of the Admiralty were 'disposed': Admiralty to Robert FitzRoy, August 1860, Appendix to the Report of the Meteorological Department of the Board of Trade, 1862; NMLA.

[169] 'As a matter of Abstract Science . . .': Robert FitzRoy to George Airy, 3 November 1860, TNA BJ 9/11, pp. 39–40.

[169] Sir John Herschel . . . backing his telegraphing: Sir John Herschel to Robert FitzRoy, 31 October 1860; Herschel Papers, Royal Society Library.

[169] 'as I am aware he occasionally theorises . . .': FitzRoy, 'On British Storms', p. 41.

[169] 'I wish his language . . .': Sir John Herschel to Robert FitzRoy, 24 October 1860; Herschel Papers, Royal Society Library.

[170] could only pit friend against friend: In the USA, Maury dedicated his *Physical Geography of the Sea* to his friend William Hasbrouck. Hasbrouck lived in Newburgh, New York, and in a coming war would be on the opposing side and therefore Maury's 'enemy'. While in mid-Atlantic on the steamer *New York* sailing back from England, Maury wrote to Hasbrouck, 'who has been such a good friend to the author from early youth to now. Till now! Do we belong to the same country yet, Hasbrouck?' (letter to 'a friend in Newburgh', 7 December 1860; quoted Corbin, *A Life of Matthew Fontaine Maury*, pp. 184–5).

[171] 'a certain degree of official notice . . .': Robert FitzRoy to Admiralty, 11 March 1861; TNA BJ 9/8, pp. 327–8.

[171] These signals are nothing to do with the Board of Trade: James Booth to the Admiralty, 23 March 1861, Appendix to the Report of the Meteorological Department of the Board of Trade, 1862; NMLA.

[171] 'expressed no inclination to proceed': *The Times*, 1 March 1861.

[172] 'I feel it a duty . . .':Robert FitzRoy to *The Times*, 2 March 1861.

[172] few wrecks occurred: Report of the Meteorological Department of the Board of Trade, 1862; NMLA.

[172] 'Not withstanding the extremely stormy weather . . .': *The Times*, 29 March 1861; in Robert FitzRoy, Report of the Meteorological Department of the Board of Trade, 1862; NMLA.

[173] 'A vessel whose destination . . .': William Maclean, Great Yarmouth; in *ibid.*

[173] the warnings 'are certainly not trusted by seafaring men . . .': Port of Shields Authority; in *ibid.*

[173] 'wedded to notions . . .': seamen of Cowes; in *ibid.*

[173] 'Notice your [weather] glasses . . .': Robert FitzRoy; in *ibid.*

[173] 'Yesterday about 3 p.m. we were warned by telegram . . .': George Hamlin to Rear-Admiral Evans, Liverpool, 22 February 1861; in *ibid.*

[174] 'Seafaring folk are proverbially averse . . .': Return from Devonport Dockyard; in *ibid.*

[174] '*General* weather probable during the next two days . . .': *The Times*, 1 August 1861.

[174] 'forecasts add almost nothing . . .': Robert FitzRoy, Report of the Meteorological Department of the Board of Trade, 1862; NMLA.

[175] 'Our system of telegraphing the weather . . .': Joseph Henry, Smithsonian Institution, to Edward Sabine, 1 July 1861; in Robert FitzRoy, *ibid.*

[175] 'on the 1st of June 1860, the first telegraphic warning by order of the Department of the Interior was given in Holland . . .': Christoph Buys Ballot, 'On the System of Forecasting the Weather Pursued in Holland', Report of the British Association for the Advancement of Science, 1863, p. 20.

[175] 'Prophesies and predictions they are not . . .': Robert FitzRoy, Report of the Meteorological Department of the Board of Trade, 1862; NMLA.

[176] 'Stones may be shaped . . .': FitzRoy, *The Weather Book*, p. 169.

CHAPTER 16. *METEOROGRAPHICA*

[177] 'As a basis to future efforts . . .': Francis Galton, circular letter to meteorological observers, July 1861; in Shaw *Manual of Meteorology*, vol. i, p. 306, fig. 119(i).

[179] 'A scientific study of the weather . . .': Francis Galton, 'Preface', in *Meteorographica: Methods of Mapping the Weather Iillustrated by 600 Diagrams of Europe during December 1861* (London and Cambridge, Macmillan & Co., 1863).

[180] 'The public have not failed to notice . . .': *The Times*, leader, 11 April 1862.

[180] 'During the last week, Nature . . .': *ibid*.

[180] 'The facts now weighed and measured mentally . . .': Robert FitzRoy, quoted in *ibid*.

[181] 'the generation now newly born . . .': *ibid*.

[181] 'there is no reason why the Admiral . . .': *ibid*.

[183] 'a branch of Meteorology . . .': Galton, 'Preface', in *Meteorographica*.

[184] 'Meteorologists are strangely behind . . .': *ibid*.

[184] 'real student of meteorology': *ibid*.

[185] 'His views about *anti*-cyclones seem . . .': Robert FitzRoy to Sir J.D. Forbes, Scottish Meteorological Society, 4 January 1864; TNA BJ 7/1045. Forbes had written to FitzRoy in praise of Galton's *Meteorographica*.

[186] He declared that, on his own behalf: Galton, 'Preface', in *Meteorographica*.

[186] 'testing the extant theory of "forecasts" . . .': *ibid*.

[187] 'The instrument does not seem to be thought practical': Robert FitzRoy to Francis Galton, 31 January 1863; TNA BJ 7/283.

CHAPTER 17. BURDEN OF PROOF

[189] 'I cannot dismiss the idea . . .': Robert FitzRoy to Sir John Herschel, 24 December 1862; Herschel Papers, Royal Society Library.

[193] According to FitzRoy, his critics: Quotations in this paragraph are taken from Robert FitzRoy, Report of the Meteorological Department of the Board of Trade, 1863; NMLA.

[195] 'Whilst powerless I heard the words "temperature" and "observation" . . .': Glaisher et al., *Travels in the Air*, pp. 53–4.

[196] 'Much has been accomplished . . .': Prospectus of the Daily Weather Map Company (1863); in Shaw, *Manual of Meteorology*, vol. i, p. 308, fig. 120(i).

[197] 'anticipate coming changes . . .': *ibid.*
[197] 'I am sorry to be obliged to say . . .': Sir John Herschel to Robert FitzRoy, 13 March 1863; Herschel Papers, Royal Society Library.

CHAPTER 18. THE FRENCH CONNECTION

[199] 'To whom these absurd but injurious predictions . . .': Robert FitzRoy to *The Times*, 18 January 1864.
[200] 'His career . . . will exhibit . . .': Washington *Daily Star*, 19 July 1861.
[202] 'I have scarcely seen Capt. Maury who is now solely a political': Robert FitzRoy to Sir John Herschel, 24 December 1862; Herschel Papers, Royal Society Library.
[204] 'I tell you these ideas freely . . .': Robert FitzRoy to M. Marié-Davy, 23 October 1863; TNA BJ 7/793.
[204] 'not so full a reply as I could desire': Robert FitzRoy to M. Marié-Davy, 4 November 1863; TNA BJ 7/795.
[204] 'as has been attempted by Mr Galton . . .': *ibid.*
[206] 'I cannot leave all theory out of sight . . .': Robert FitzRoy to Matteucci, 2 April 1864; in Report of the Meteorological Department of the Board of Trade, 1864, Appendix 1, NMLA.
[209] 'I have just read the Bulletin . . .': Robert FitzRoy to Thomas Babington, memo, 24 February 1864; TNA BJ 7/45.
[209] 'disadvantage of forecasting in an insular position': Matthew Maury to Robert FitzRoy, 30 February 1864. The phrase is written against an entry in the 'Register of Foreign or Scientific Letters Written and Received (in brief)', TNA BJ 9/12, but the letter itself is missing.
[209] 'Nothing could now be more fallacious': Report of the Meteorological Department, 1864; NMLA.
[210] 'Only in consequence of attentive . . .': *ibid.*
[210] 'In discouraging such forecasts . . .': *ibid.*

CHAPTER 19. FORECASTING THE END

[212] 'This is the third time I have cautioned . . .': Robert FitzRoy to Thomas Babington, memo, 11 February 1864; Met Office internal memos February–May 1864, TNA BJ 7/45.
[212] 'I am exceedingly vexed at . . .': Robert FitzRoy to Thomas Babington, 9 May 1864; *ibid.*
[213] 'a morbid depression of spirits': Charles Darwin to his sister Catherine, from Valparaiso, 8 November 1834; Francis Darwin (ed.), *Life and Letters of Charles Darwin*, vol. i, pp. 228–9.

[213] Mr Augustus Smith, MP for Truro, was immediately on his feet: A report of Mr Smith's speech was published in the parliamentary section of *The Times*, 13 May 1864.

[214] FitzRoy . . . issued a detailed defence, which appeared next day in *The Times*: Robert FitzRoy to *The Times*, 14 May 1864.

[214] 'A separate official analysis and digest . . .': Robert FitzRoy, Report of the Meteorological Department of the Board of Trade, 1864; NMLA.

[215] 'too much praise cannot be given to the author . . .': Robert Eggert, Harbour Master, King's Lynn, to the Editor of the *Lynn Advertiser*, 31 December 1863; in Report of the Meteorological Department of the Board of Trade, 1864; NMLA.

[215] 'Whatever may be the progress of science . . .': This and the following quotations are taken from *The Times*, leader, 18 June 1864.

[218] 'being hard of hearing at any time': Robert FitzRoy to Thomas Babington, memo, 6 May 1864; TNA BJ 7/755.

[218] 'his naturally clear and vigorous mind . . .': Thomas Babington to Edward Sabine, report, October 1867; TNA BJ 7/872.

[220] 'Yesterday morning a painful feeling of regret . . .': *The Times*, 2 May 1865.

[220] 'a circumstance which had impressed him awfully': from a description of the days leading up to the death of FitzRoy, written by his widow; private papers in the possession of the descendants of Robert FitzRoy, quoted in Mellersh, *FitzRoy of the Beagle*, p. 283.

[223] 'He wished to know what was the intention . . .': Sir Augustus Smith, speech, parliamentary report, *The Times*, 9 June 1865.

[223] 'In the opinion of those best qualified to judge . . .': Milner Gibson, reply in the House to Sir Augustus Smith, parliamentary report, *The Times*, 9 June 1865.

[223] 'had been done by the late Admiral . . .': *ibid.*

[223] 'Since that Admiral's lamented death . . .': *ibid.*

[224] 'We can find no competant meteorologist . . .': Report of the Royal Society's Committee of Enquiry, 1866; NMLA.

[228] '[Le Verrier] abandons the notion . . .': Farrer to Gibson, 29 April 1866; TNA BJ 7/960.

[228] 'Very interesting – but after all . . .': Milner Gibson to Thomas Farrer, copied to the Royal Society, Captain Evans and Francis Galton, 17 May 1866; TNA BJ 7/960.

[229] 'digestion and tabulation of results': Report of the Royal Society's Committee of Enquiry, 1866; NMLA.

[229] 'At present, these warnings are founded . . .': Reply from the President and Council of the Royal Society on the report of the

Committee of Enquiry, October 1866, quoted in Appendix II of the Report of the Meteorological Committee of the Royal Society, 1867.

[229] 'Am I not right in supposing . . .': Thomas Farrer to Edward Sabine, 18 November 1866; TNA BJ 7/873.

[229] On 29 November 1866 the Board of Trade issued a circular announcing that cautionary storm warnings would be discontinued as from 7 December: A draft is filed in TNA BJ 7/874.

CHAPTER 20. REBIRTH

[230] 'I trust the suspension will not be of long continuance . . .': Revd Francis Redford, living in St Paul's Parsonage at Silloth on the Cumbrian coast, who was himself a Fellow of the Royal Society, to the Board of Trade; PP (1867) LXIV, pp. 185–203.

[230] 'but surely this furnishes no valid reason . . .': Memorial from the Dundee Harbour Trustees, 10 December 1866; PP (1867) LXIV, pp. 185–203.

[230] 'the extreme disappointment . . .': *ibid.*

[231] 'set a high value . . .': petition from the Port of Leith; PP (1867) LXIV, pp. 185–203.

[231] 'The Local Board would not be understood . . .': Dundee Local Marine Board to the Board of Trade, 21 January 1867; PP (1867) LXIV, pp. 185–203.

[231] Liverpool Underwriters were 'much in favour of the cautionary signals': Liverpool Underwriters Association; PP (1867) LXIV, pp. 185–203.

[231] 'however imperfect the "Storm Signals" were . . .': Underwriters and shipowners of Glasgow and Greenock; PP (1867) LXIV, pp. 185–203.

[231] 'in one of our largest collieries . . .': Board of Directors of the Manchester Chamber of Commerce, 10 January 1867; PP (1867) LXIV, pp. 185–203.

[231] 'Now, it is quite true, that . . .': C. Piazzi Smyth, Astronomer Royal for Scotland, to the Edinburgh Chamber of Commerce, to be passed on to the Board of Trade, 6 January 1867; PP (1867) LXIV, pp. 185–203.

[233] 'A large deputation has waited on the Duke of Richmond . . .': Board of Trade to Meteorological Committee (Thomas Farrer to Robert Scott), 31 May 1867; in Report of the Meteorological Committee of the Royal Society, 1867; NMLA.

[234] 'it must not be forgotten that storm warnings . . .': Reply from the President and Council of the Royal Society, 27 October

1866, quoted in Appendix II of the Report of the Meteorological Committee of the Royal Society, 1867; NMLA.

[234] 'we might have prevented and altered . . .': Thomas Farrer to Edward Sabine, 18 November 1866, TNA BJ 7/873.

[234] In a letter to Farrer in April 1867: Edward Sabine to Thomas Farrer, 20/21 April 1867; TNA BJ 7/878.

[234] he was complaining that the Treasury: Edward Sabine to Thomas Farrer, 23 April 1867; TNA BJ 7/879.

[237] the committee declined to 'prognosticate weather . . .': Meteorological Committee to Thomas Farrer, 8 June 1867; PP (1867) LXIV, p. 209.

[237] At the end of October, the committee received a letter: Board of Trade to Meteorological Committee, 30 October 1867, enclosing a letter from Elphinstone to Duke of Richmond; in Report of the Meteorological Committee of the Royal Society, 1867; NMLA.

[238] 'Storm west of Penzance . . .': Robert Scott, on behalf of the Meteorological Committee of the Royal Society to the Board of Trade, 8 June 1867; PP (1867) LXIV, pp. 185–203.

[238] 'With a view of collection and distributing . . .': *ibid.*

[239] 'fortunately these shores are little frequented by shipping': Report of the Meteorological Committee of the Royal Society, 1868; NMLA.

CHAPTER 21. A NEW DIMENSION

[247] 'The time is coming . . .': Matthew Maury, address delivered in St Louis, Griffin, Norfolk, Richmond and Nashville, USA, 1871; in Corbin, *A Life of Matthew Fontaine Maury*, p. 280.

[257] 'Does the flap of a butterfly's wings . . .': the title of a paper presented by Edward Lorenz in 1972, quoted in Ziauddin Sardar and Iwona Abrams, *Introducing Chaos* (Cambridge, Icon Books, 1999), p. 54. From this paper originated the phrase 'The Butterfly Effect', symbolic of Chaos Theory and popularised by James Gleick in his 1988 book, *Chaos*.

[258] 'It is not necessary to impress upon a people . . .': *The Times*, 11 April 1862.

Bibliography

Ashley, Bernard, *Weather Man*, London, Allman & Son, 1970

Audric, Brian, 'The Meteorological Office Dunstable and the IDA Unit in World War II' , Occasional Papers on Meteorological History, No. 2, Reading, The Royal Meteorological Society Specialist Group for the History of Meteorology and Physical Oceanography (September 2000)

Barlow, Derek, 'From Wind Stars to Weather Forecasts: The Last Voyage of Admiral Robert FitzRoy', *Weather*, 49/4 (April 1994)

Barlow, Derek, 'The Devil Within: Evolution of a Tragedy', *Weather*, 52/11 (November 1997)

—— 'Origins of Meteorology', a catalogue of correspondence and papers relating to the first Meteorological Office under FitzRoy (1854–65), Babington (1865–6) and the first two years of the Meteorological Office under the Scientific Committee of the Royal Society, National Archives

Blench, Brian J.R., 'Luke Howard and his Contribution to Meteorology', *Weather*, 18/3 (March 1963)

Burton, Jim, 'Robert FitzRoy and the Early History of the Meteorological Office', *British Journal for the History of Science*, 19 (1986), pp. 147–76

—— 'Pen Portraits of Presidents – Robert Henry Scott, MA, Dsc, FRS', *Weather*, 49/11 (November 1994)

—— 'Pen Portraits of Presidents – Sir Napier Shaw, MA, ScD, LLD, FRS', *Weather*, 50/3 (March 1995)

Buys Ballot, Christoph, 'On the System of Forecasting the Weather Pursued in Holland', Report of the British Association for the Advancement of Science, 1863

Caskie, Jaquelin Ambler, *Life and Letters of Matthew Fontaine Maury*, Richmond, Va., Richmond Press, 1928

Cooke, Christopher, *FitzRoy's Facts and Failures*, London, Hall, 1867

Corbin, Diana Fontaine Maury, *A Life of Matthew Fontaine Maury*, London, Sampson Low, Marston, Searle & Rivington, 1888

Courtney, Nicholas, *Gale Force 10: The Life and Legacy of Admiral Beaufort 1774–1857*, London, Review, 2002

Crewe, M.E., 'Meteorology and Aerial Navigation' , Occasional Papers on Meteorological History, No. 4, Reading, The Royal Meteorological Society Specialist Group for the History of Meteorology and Physical Oceanography (September 2002)

Dalton, John, 'Experimental Enquiry into the Proportion of the Several Gases or Elastic Fluids, Constituting the Atmosphere', *Memoirs of the Literary and Philosophical Society of Manchester 2 series 1, 1799–1805* (1805), pp. 244–58; read 12 November 1802

Darwin, Francis (ed.), *Life and Letters of Charles Darwin*, 4 vols, London, John Murray, 1887

—— and Seward, A.C. (eds), *More Letters of Charles Darwin*, 2 vols, London, John Murray, 1903

Davis, John L., 'Weather Forecasting and the Development of Meteorological Theory at the Paris Observatory 1853–78', *Annals of Science*, 41 (1984), pp. 359–82

Dugdale, George Stratford, *Whitehall through the Centuries*, London, Phoenix House, 1950

Espy, James P., 'On Storms', Report of the British Association for the Advancement of Science, 1840

Falk, Bernard, *The Royal Fitz Roys: Dukes of Grafton through Four Centuries*, London, Hutchinson, 1950

Field, M., 'Meteorologists Profile – Charles Kenneth MacKinnon Douglas, OBE, AFC, MA', *Weather*, 54/10 (October 1999)

FitzRoy, Robert, *A Narrative of the Surveying Voyages of His Majesty's Ships Adventure and Beagle Volume II*, London, Henry Colburn, 1839

—— 'Wind-charts of the Atlantic, compiled from Maury's Pilot Charts', Report of the British Association for the Advancement of Science, 1855

—— *Marine Barometer Adopted by HM Government and the US Navy with Method of Testing Marine Meteorological Instruments*, Board of Trade Miscellaneous Paper, London, HMSO, 1858

—— *Barometer and Weather Guide*, Board of Trade, 1858; 2nd edn, London, Eyre & Spottiswoode, 1859

—— *Notes on Meteorology*, Board of Trade Miscellaneous Paper, London, HMSO, 1859

—— 'On British Storms', Report of the British Association for the Advancement of Science, 1860, pp. 39–44

—— *The Weather Book: A Manual of Practical Meteorology*, London, Longman, Green, Longman, Roberts & Green, 1863

—— Report of the Meteorological Office, 1855, unpublished

—— Reports of the Meteorological Office of the Board of Trade for the years 1857, 1858, 1862, 1863, 1864, all issued by the Board of Trade

Fleming, James Rodger, *Meteorology in America 1800–1870*, Baltimore, Johns Hopkins University Press, 1980

Forbes, James D., 'Report upon the Recent Progress and Present State of Meteorology', Second Report of the British Association for the Advancement of Science, 1832

Foreman, Susan, *Shoes and Ships and Sealing Wax: Illustrated History of the Board of Trade 1786–1986*, London, Department of Trade and Industry, HMSO, 1986

Galton, Francis, *Meteorographica: Methods of Mapping the Weather Illustrated by 600 Diagrams of Europe during December 1861*, London and Cambridge, Macmillan & Co., 1863

Glaisher, James, Flammarion, Camille, de Fonvielle, W., and Tissandier, Gaston, *Travels in the Air*, ed. James Glaisher, London, Richard Bentley, 1871

Hearn, Chester G., *Tracks in the Sea: Matthew Fontaine Maury and the Mapping of the Oceans*, London, International Marine/McGraw-Hill, 2002

Heath, Admiral Sir Leopold George, KCB, *Letters from the Black Sea during the Crimean War 1854–55*, London, Bentley, 1897

Hill, J.R. (ed.), *The Oxford Illustrated History of the Royal Navy*, Oxford, Oxford University Press, 1995

Howard, Luke, *Essay on the Modifications of Clouds*, 1st edn, 1804; repr. London, 1830

Hunting, Penelope, *Royal Westminster*, London, Royal Institute of Chartered Surveyors, 1981

Jahns, Patricia, *Matthew Fontaine Maury and Joseph Henry: Scientists of the Civil War*, New York, Hastings House, 1961

Jankovic, Vladimir, *Reading the Skies: A Cultural History of English Weather 1650–1820*, Manchester, Manchester University Press, 2000

Judd, Dennis, *The Crimean War*, London, Hart-Davis MacGibbon, 1975

Kelly, Mrs Tom, *From the Fleet in the Fifties*, incorporating letters by the Revd S. Kelson Stothert, Chaplain to the Naval Brigade, London, Hurst & Blackett Ltd, 1902

Knowles Middleton, W.E., *A History of the Barometer*, Baltimore, Johns Hopkins University Press, 1964

—— *A History of the Theories of Rain and Other Forms of Precipitation*, London, Oldbourne, 1965

—— *A History of the Thermometer and its Uses in Meteorology*, Baltimore, Johns Hopkins University Press, 1966

—— and Spilhaus, A.F., *Meteorological Instruments*, Toronto, University of Toronto Press, 1953

Lewis, R.P.W., 'The Founding of the Meteorological Office 1854–55', *Meteorological Magazine*, 110 (1981)

Mellersh, H.E.L., *FitzRoy of the Beagle*, London, Rupert Hart-Davis, 1968

Palmer, Alan, *The Banner of Battle: The Story of the Crimean War*, London, Weidenfeld & Nicolson, 1987

Pearson, Karl, *Life, Letters and Labours of Francis Galton*, Cambridge, Cambridge University Press, 1914

Pedgley, D.E., 'Pen Portraits of Presidents – James Glaisher, FRS', *Weather*, 50/11 (November 1995)

Reid, Lt-Col William, 'A Statement of the Progress made towards developing the Law of Storms; and of what seems further desirable to be done, to the advance of our knowledge of the subject', Eighth Report of the British Association for the Advancement of Science, 1838

Robins, Major Colin (ed.), *Captain Dunscombe's Diary: The Diary of Captain Nicholas Dunscombe, 46th (South Devonshire) Foot*, Bowdon, Withycut House, 2003

Sardar, Ziauddin, and Abrams, Iwona, *Introducing Chaos*, Cambridge, Icon Books, 1999

Scott, Douglas Frederick Schumacher, *Luke Howard: His Correspondence with Goethe and his Continental Journey of 1816*, York, Williams Sessions, 1976

Shaw, Sir W. Napier, *Manual of Meteorology*, i. *Meteorology in History*, Cambridge, Cambridge University Press, 1926

Stanbury, David (ed.), *A Narrative of the Voyage of HMS Beagle*, London, Folio Society, 1977

Wayland, John Walker, *Pathfinder of the Seas: The Life of Matthew Fontaine Maury*, Richmond, Va., Garrett & Maisie, 1930

Williams, James Thaxter, *The History of the Weather*, New York, Nova Science Publishers, 1999

World Meteorological Organisation, *International Cloud Atlas Vol. 1*, 1939

Wrottesley, Baron John, 'On Lt Maury's plan for improving navigation, with some remarks upon the advantages arising from the pursuit of abstract science', text of a speech presented in the House of Lords, London, 1853

Index